Planetary Systems: A Very Short Introduction

VERY SHORT INTRODUCTIONS are for anyone wanting a stimulating and accessible way into a new subject. They are written by experts, and have been translated into more than 45 different languages.

The series began in 1995, and now covers a wide variety of topics in every discipline. The VSI library currently contains over 700 volumes—a Very Short Introduction to everything from Psychology and Philosophy of Science to American History and Relativity—and continues to grow in every subject area.

Very Short Introductions available now:

ABOLITIONISM Richard S. Newman
THE ABRAHAMIC RELIGIONS
    Charles L. Cohen
ACCOUNTING Christopher Nobes
ADAM SMITH Christopher J. Berry
ADOLESCENCE Peter K. Smith
ADVERTISING Winston Fletcher
AERIAL WARFARE Frank Ledwidge
AESTHETICS Bence Nanay
AFRICAN AMERICAN RELIGION
    Eddie S. Glaude Jr
AFRICAN HISTORY John Parker and
    Richard Rathbone
AFRICAN POLITICS Ian Taylor
AFRICAN RELIGIONS
    Jacob K. Olupona
AGEING Nancy A. Pachana
AGNOSTICISM Robin Le Poidevin
AGRICULTURE Paul Brassley and
    Richard Soffe
ALBERT CAMUS Oliver Gloag
ALEXANDER THE GREAT
    Hugh Bowden
ALGEBRA Peter M. Higgins
AMERICAN CULTURAL HISTORY
    Eric Avila
AMERICAN FOREIGN RELATIONS
    Andrew Preston
AMERICAN HISTORY Paul S. Boyer
AMERICAN IMMIGRATION
    David A. Gerber
AMERICAN LEGAL HISTORY
    G. Edward White
AMERICAN NAVAL HISTORY
    Craig L. Symonds

AMERICAN POLITICAL HISTORY
    Donald Critchlow
AMERICAN POLITICAL PARTIES
    AND ELECTIONS L. Sandy Maisel
AMERICAN POLITICS
    Richard M. Valelly
THE AMERICAN PRESIDENCY
    Charles O. Jones
THE AMERICAN REVOLUTION
    Robert J. Allison
AMERICAN SLAVERY
    Heather Andrea Williams
THE AMERICAN WEST Stephen Aron
AMERICAN WOMEN'S HISTORY
    Susan Ware
ANAESTHESIA Aidan O'Donnell
ANALYTIC PHILOSOPHY
    Michael Beaney
ANARCHISM Colin Ward
ANCIENT ASSYRIA Karen Radner
ANCIENT EGYPT Ian Shaw
ANCIENT EGYPTIAN ART AND
    ARCHITECTURE Christina Riggs
ANCIENT GREECE Paul Cartledge
THE ANCIENT NEAR EAST
    Amanda H. Podany
ANCIENT PHILOSOPHY Julia Annas
ANCIENT WARFARE
    Harry Sidebottom
ANGELS David Albert Jones
ANGLICANISM Mark Chapman
THE ANGLO-SAXON AGE
    John Blair
ANIMAL BEHAVIOUR
    Tristram D. Wyatt

THE ANIMAL KINGDOM
    Peter Holland
ANIMAL RIGHTS David DeGrazia
THE ANTARCTIC Klaus Dodds
ANTHROPOCENE Erle C. Ellis
ANTISEMITISM Steven Beller
ANXIETY Daniel Freeman and
    Jason Freeman
THE APOCRYPHAL GOSPELS
    Paul Foster
APPLIED MATHEMATICS
    Alain Goriely
ARCHAEOLOGY Paul Bahn
ARCHITECTURE Andrew Ballantyne
ARISTOCRACY William Doyle
ARISTOTLE Jonathan Barnes
ART HISTORY Dana Arnold
ART THEORY Cynthia Freeland
ARTIFICIAL INTELLIGENCE
    Margaret A. Boden
ASIAN AMERICAN HISTORY
    Madeline Y. Hsu
ASTROBIOLOGY David C. Catling
ASTROPHYSICS James Binney
ATHEISM Julian Baggini
THE ATMOSPHERE Paul I. Palmer
AUGUSTINE Henry Chadwick
AUSTRALIA Kenneth Morgan
AUTISM Uta Frith
AUTOBIOGRAPHY Laura Marcus
THE AVANT GARDE David Cottington
THE AZTECS David Carrasco
BABYLONIA Trevor Bryce
BACTERIA Sebastian G. B. Amyes
BANKING John Goddard and
    John O. S. Wilson
BARTHES Jonathan Culler
THE BEATS David Sterritt
BEAUTY Roger Scruton
BEHAVIOURAL ECONOMICS
    Michelle Baddeley
BESTSELLERS John Sutherland
THE BIBLE John Riches
BIBLICAL ARCHAEOLOGY
    Eric H. Cline
BIG DATA Dawn E. Holmes
BIOGRAPHY Hermione Lee
BIOMETRICS Michael Fairhurst
BLACK HOLES Katherine Blundell
BLOOD Chris Cooper
THE BLUES Elijah Wald

THE BODY Chris Shilling
THE BOOK OF COMMON PRAYER
    Brian Cummings
THE BOOK OF MORMON
    Terryl Givens
BORDERS Alexander C. Diener and
    Joshua Hagen
THE BRAIN Michael O'Shea
BRANDING Robert Jones
THE BRICS Andrew F. Cooper
THE BRITISH CONSTITUTION
    Martin Loughlin
THE BRITISH EMPIRE Ashley Jackson
BRITISH POLITICS Anthony Wright
BUDDHA Michael Carrithers
BUDDHISM Damien Keown
BUDDHIST ETHICS Damien Keown
BYZANTIUM Peter Sarris
C. S. LEWIS James Como
CALVINISM Jon Balserak
CANCER Nicholas James
CAPITALISM James Fulcher
CATHOLICISM Gerald O'Collins
CAUSATION Stephen Mumford and
    Rani Lill Anjum
THE CELL Terence Allen and
    Graham Cowling
THE CELTS Barry Cunliffe
CHAOS Leonard Smith
CHARLES DICKENS Jenny Hartley
CHEMISTRY Peter Atkins
CHILD PSYCHOLOGY Usha Goswami
CHILDREN'S LITERATURE
    Kimberley Reynolds
CHINESE LITERATURE Sabina Knight
CHOICE THEORY Michael Allingham
CHRISTIAN ART Beth Williamson
CHRISTIAN ETHICS D. Stephen Long
CHRISTIANITY Linda Woodhead
CIRCADIAN RHYTHMS
    Russell Foster and Leon Kreitzman
CITIZENSHIP Richard Bellamy
CIVIL ENGINEERING
    David Muir Wood
CLASSICAL LITERATURE William Allan
CLASSICAL MYTHOLOGY
    Helen Morales
CLASSICS Mary Beard and
    John Henderson
CLAUSEWITZ Michael Howard
CLIMATE Mark Maslin

CLIMATE CHANGE Mark Maslin
CLINICAL PSYCHOLOGY
    Susan Llewelyn and
    Katie Aafjes-van Doorn
COGNITIVE NEUROSCIENCE
    Richard Passingham
THE COLD WAR Robert McMahon
COLONIAL AMERICA Alan Taylor
COLONIAL LATIN AMERICAN
    LITERATURE Rolena Adorno
COMBINATORICS Robin Wilson
COMEDY Matthew Bevis
COMMUNISM Leslie Holmes
COMPARATIVE LITERATURE
    Ben Hutchinson
COMPLEXITY John H. Holland
THE COMPUTER Darrel Ince
COMPUTER SCIENCE
    Subrata Dasgupta
CONCENTRATION CAMPS Dan Stone
CONFUCIANISM Daniel K. Gardner
THE CONQUISTADORS
    Matthew Restall and
    Felipe Fernández-Armesto
CONSCIENCE Paul Strohm
CONSCIOUSNESS Susan Blackmore
CONTEMPORARY ART
    Julian Stallabrass
CONTEMPORARY FICTION
    Robert Eaglestone
CONTINENTAL PHILOSOPHY
    Simon Critchley
COPERNICUS Owen Gingerich
CORAL REEFS Charles Sheppard
CORPORATE SOCIAL
    RESPONSIBILITY Jeremy Moon
CORRUPTION Leslie Holmes
COSMOLOGY Peter Coles
COUNTRY MUSIC Richard Carlin
CRIME FICTION Richard Bradford
CRIMINAL JUSTICE Julian V. Roberts
CRIMINOLOGY Tim Newburn
CRITICAL THEORY
    Stephen Eric Bronner
THE CRUSADES Christopher Tyerman
CRYPTOGRAPHY Fred Piper and
    Sean Murphy
CRYSTALLOGRAPHY A. M. Glazer
THE CULTURAL REVOLUTION
    Richard Curt Kraus

DADA AND SURREALISM
    David Hopkins
DANTE Peter Hainsworth and
    David Robey
DARWIN Jonathan Howard
THE DEAD SEA SCROLLS
    Timothy H. Lim
DECADENCE David Weir
DECOLONIZATION Dane Kennedy
DEMOCRACY Bernard Crick
DEMOGRAPHY Sarah Harper
DEPRESSION Jan Scott and
    Mary Jane Tacchi
DERRIDA Simon Glendinning
DESCARTES Tom Sorell
DESERTS Nick Middleton
DESIGN John Heskett
DEVELOPMENT Ian Goldin
DEVELOPMENTAL BIOLOGY
    Lewis Wolpert
THE DEVIL Darren Oldridge
DIASPORA Kevin Kenny
DICTIONARIES Lynda Mugglestone
DINOSAURS David Norman
DIPLOMACY Joseph M. Siracusa
DOCUMENTARY FILM
    Patricia Aufderheide
DREAMING J. Allan Hobson
DRUGS Les Iversen
DRUIDS Barry Cunliffe
DYNASTY Jeroen Duindam
DYSLEXIA Margaret J. Snowling
EARLY MUSIC Thomas Forrest Kelly
THE EARTH Martin Redfern
EARTH SYSTEM SCIENCE Tim Lenton
ECONOMICS Partha Dasgupta
EDUCATION Gary Thomas
EGYPTIAN MYTH Geraldine Pinch
EIGHTEENTH–CENTURY BRITAIN
    Paul Langford
THE ELEMENTS Philip Ball
EMOTION Dylan Evans
EMPIRE Stephen Howe
ENERGY SYSTEMS Nick Jenkins
ENGELS Terrell Carver
ENGINEERING David Blockley
THE ENGLISH LANGUAGE
    Simon Horobin
ENGLISH LITERATURE
    Jonathan Bate

THE ENLIGHTENMENT John Robertson
ENTREPRENEURSHIP Paul Westhead and Mike Wright
ENVIRONMENTAL ECONOMICS Stephen Smith
ENVIRONMENTAL ETHICS Robin Attfield
ENVIRONMENTAL LAW Elizabeth Fisher
ENVIRONMENTAL POLITICS Andrew Dobson
EPICUREANISM Catherine Wilson
EPIDEMIOLOGY Rodolfo Saracci
ETHICS Simon Blackburn
ETHNOMUSICOLOGY Timothy Rice
THE ETRUSCANS Christopher Smith
EUGENICS Philippa Levine
THE EUROPEAN UNION Simon Usherwood and John Pinder
EUROPEAN UNION LAW Anthony Arnull
EVOLUTION Brian and Deborah Charlesworth
EXISTENTIALISM Thomas Flynn
EXPLORATION Stewart A. Weaver
EXTINCTION Paul B. Wignall
THE EYE Michael Land
FAIRY TALE Marina Warner
FAMILY LAW Jonathan Herring
FASCISM Kevin Passmore
FASHION Rebecca Arnold
FEDERALISM Mark J. Rozell and Clyde Wilcox
FEMINISM Margaret Walters
FILM Michael Wood
FILM MUSIC Kathryn Kalinak
FILM NOIR James Naremore
THE FIRST WORLD WAR Michael Howard
FOLK MUSIC Mark Slobin
FOOD John Krebs
FORENSIC PSYCHOLOGY David Canter
FORENSIC SCIENCE Jim Fraser
FORESTS Jaboury Ghazoul
FOSSILS Keith Thomson
FOUCAULT Gary Gutting
THE FOUNDING FATHERS R. B. Bernstein

FRACTALS Kenneth Falconer
FREE SPEECH Nigel Warburton
FREE WILL Thomas Pink
FREEMASONRY Andreas Önnerfors
FRENCH LITERATURE John D. Lyons
THE FRENCH REVOLUTION William Doyle
FREUD Anthony Storr
FUNDAMENTALISM Malise Ruthven
FUNGI Nicholas P. Money
THE FUTURE Jennifer M. Gidley
GALAXIES John Gribbin
GALILEO Stillman Drake
GAME THEORY Ken Binmore
GANDHI Bhikhu Parekh
GARDEN HISTORY Gordon Campbell
GENES Jonathan Slack
GENIUS Andrew Robinson
GENOMICS John Archibald
GEOFFREY CHAUCER David Wallace
GEOGRAPHY John Matthews and David Herbert
GEOLOGY Jan Zalasiewicz
GEOPHYSICS William Lowrie
GEOPOLITICS Klaus Dodds
GERMAN LITERATURE Nicholas Boyle
GERMAN PHILOSOPHY Andrew Bowie
GLACIATION David J. A. Evans
GLOBAL CATASTROPHES Bill McGuire
GLOBAL ECONOMIC HISTORY Robert C. Allen
GLOBALIZATION Manfred Steger
GOD John Bowker
GOETHE Ritchie Robertson
THE GOTHIC Nick Groom
GOVERNANCE Mark Bevir
GRAVITY Timothy Clifton
THE GREAT DEPRESSION AND THE NEW DEAL Eric Rauchway
HABERMAS James Gordon Finlayson
THE HABSBURG EMPIRE Martyn Rady
HAPPINESS Daniel M. Haybron
THE HARLEM RENAISSANCE Cheryl A. Wall
THE HEBREW BIBLE AS LITERATURE Tod Linafelt
HEGEL Peter Singer
HEIDEGGER Michael Inwood

THE HELLENISTIC AGE
  Peter Thonemann
HEREDITY John Waller
HERMENEUTICS Jens Zimmermann
HERODOTUS Jennifer T. Roberts
HIEROGLYPHS Penelope Wilson
HINDUISM Kim Knott
HISTORY John H. Arnold
THE HISTORY OF ASTRONOMY
  Michael Hoskin
THE HISTORY OF CHEMISTRY
  William H. Brock
THE HISTORY OF CHILDHOOD
  James Marten
THE HISTORY OF CINEMA
  Geoffrey Nowell-Smith
THE HISTORY OF LIFE
  Michael Benton
THE HISTORY OF MATHEMATICS
  Jacqueline Stedall
THE HISTORY OF MEDICINE
  William Bynum
THE HISTORY OF PHYSICS
  J. L. Heilbron
THE HISTORY OF TIME
  Leofranc Holford-Strevens
HIV AND AIDS Alan Whiteside
HOBBES Richard Tuck
HOLLYWOOD Peter Decherney
THE HOLY ROMAN EMPIRE
  Joachim Whaley
HOME Michael Allen Fox
HOMER Barbara Graziosi
HORMONES Martin Luck
HUMAN ANATOMY
  Leslie Klenerman
HUMAN EVOLUTION Bernard Wood
HUMAN RIGHTS Andrew Clapham
HUMANISM Stephen Law
HUME A. J. Ayer
HUMOUR Noël Carroll
THE ICE AGE Jamie Woodward
IDENTITY Florian Coulmas
IDEOLOGY Michael Freeden
THE IMMUNE SYSTEM
  Paul Klenerman
INDIAN CINEMA
  Ashish Rajadhyaksha
INDIAN PHILOSOPHY Sue Hamilton
THE INDUSTRIAL REVOLUTION
  Robert C. Allen

INFECTIOUS DISEASE
  Marta L. Wayne
  and Benjamin M. Bolker
INFINITY Ian Stewart
INFORMATION Luciano Floridi
INNOVATION Mark Dodgson and
  David Gann
INTELLECTUAL PROPERTY
  Siva Vaidhyanathan
INTELLIGENCE Ian J. Deary
INTERNATIONAL LAW
  Vaughan Lowe
INTERNATIONAL MIGRATION
  Khalid Koser
INTERNATIONAL RELATIONS
  Paul Wilkinson
INTERNATIONAL SECURITY
  Christopher S. Browning
IRAN Ali M. Ansari
ISLAM Malise Ruthven
ISLAMIC HISTORY Adam Silverstein
ISOTOPES Rob Ellam
ITALIAN LITERATURE
  Peter Hainsworth and David Robey
JESUS Richard Bauckham
JEWISH HISTORY David N. Myers
JOURNALISM Ian Hargreaves
JUDAISM Norman Solomon
JUNG Anthony Stevens
KABBALAH Joseph Dan
KAFKA Ritchie Robertson
KANT Roger Scruton
KEYNES Robert Skidelsky
KIERKEGAARD Patrick Gardiner
KNOWLEDGE Jennifer Nagel
THE KORAN Michael Cook
KOREA Michael J. Seth
LAKES Warwick F. Vincent
LANDSCAPE ARCHITECTURE
  Ian H. Thompson
LANDSCAPES AND
  GEOMORPHOLOGY
  Andrew Goudie and Heather Viles
LANGUAGES Stephen R. Anderson
LATE ANTIQUITY Gillian Clark
LAW Raymond Wacks
THE LAWS OF THERMODYNAMICS
  Peter Atkins
LEADERSHIP Keith Grint
LEARNING Mark Haselgrove
LEIBNIZ Maria Rosa Antognazza

LEO TOLSTOY Liza Knapp
LIBERALISM Michael Freeden
LIGHT Ian Walmsley
LINCOLN Allen C. Guelzo
LINGUISTICS Peter Matthews
LITERARY THEORY Jonathan Culler
LOCKE John Dunn
LOGIC Graham Priest
LOVE Ronald de Sousa
MACHIAVELLI Quentin Skinner
MADNESS Andrew Scull
MAGIC Owen Davies
MAGNA CARTA Nicholas Vincent
MAGNETISM Stephen Blundell
MALTHUS Donald Winch
MAMMALS T. S. Kemp
MANAGEMENT John Hendry
MAO Delia Davin
MARINE BIOLOGY Philip V. Mladenov
THE MARQUIS DE SADE
    John Phillips
MARTIN LUTHER Scott H. Hendrix
MARTYRDOM Jolyon Mitchell
MARX Peter Singer
MATERIALS Christopher Hall
MATHEMATICAL FINANCE
    Mark H. A. Davis
MATHEMATICS Timothy Gowers
MATTER Geoff Cottrell
THE MEANING OF LIFE
    Terry Eagleton
MEASUREMENT David Hand
MEDICAL ETHICS Michael Dunn and
    Tony Hope
MEDICAL LAW Charles Foster
MEDIEVAL BRITAIN John Gillingham
    and Ralph A. Griffiths
MEDIEVAL LITERATURE
    Elaine Treharne
MEDIEVAL PHILOSOPHY
    John Marenbon
MEMORY Jonathan K. Foster
METAPHYSICS Stephen Mumford
METHODISM William J. Abraham
THE MEXICAN REVOLUTION
    Alan Knight
MICHAEL FARADAY
    Frank A. J. L. James
MICROBIOLOGY Nicholas P. Money
MICROECONOMICS Avinash Dixit
MICROSCOPY Terence Allen

THE MIDDLE AGES Miri Rubin
MILITARY JUSTICE Eugene R. Fidell
MILITARY STRATEGY
    Antulio J. Echevarria II
MINERALS David Vaughan
MIRACLES Yujin Nagasawa
MODERN ARCHITECTURE
    Adam Sharr
MODERN ART David Cottington
MODERN CHINA Rana Mitter
MODERN DRAMA
    Kirsten E. Shepherd-Barr
MODERN FRANCE
    Vanessa R. Schwartz
MODERN INDIA Craig Jeffrey
MODERN IRELAND Senia Pašeta
MODERN ITALY Anna Cento Bull
MODERN JAPAN
    Christopher Goto-Jones
MODERN LATIN AMERICAN
    LITERATURE
    Roberto González Echevarría
MODERN WAR Richard English
MODERNISM Christopher Butler
MOLECULAR BIOLOGY Aysha Divan
    and Janice A. Royds
MOLECULES Philip Ball
MONASTICISM Stephen J. Davis
THE MONGOLS Morris Rossabi
MOONS David A. Rothery
MORMONISM
    Richard Lyman Bushman
MOUNTAINS Martin F. Price
MUHAMMAD Jonathan A. C. Brown
MULTICULTURALISM Ali Rattansi
MULTILINGUALISM John C. Maher
MUSIC Nicholas Cook
MYTH Robert A. Segal
NAPOLEON David Bell
THE NAPOLEONIC WARS
    Mike Rapport
NATIONALISM Steven Grosby
NATIVE AMERICAN LITERATURE
    Sean Teuton
NAVIGATION Jim Bennett
NAZI GERMANY Jane Caplan
NELSON MANDELA Elleke Boehmer
NEOLIBERALISM Manfred Steger and
    Ravi Roy
NETWORKS Guido Caldarelli and
    Michele Catanzaro

THE NEW TESTAMENT
  Luke Timothy Johnson
THE NEW TESTAMENT AS
  LITERATURE Kyle Keefer
NEWTON Robert Iliffe
NIELS BOHR J. L. Heilbron
NIETZSCHE Michael Tanner
NINETEENTH–CENTURY BRITAIN
  Christopher Harvie and
  H. C. G. Matthew
THE NORMAN CONQUEST
  George Garnett
NORTH AMERICAN INDIANS
  Theda Perdue and Michael D. Green
NORTHERN IRELAND
  Marc Mulholland
NOTHING Frank Close
NUCLEAR PHYSICS Frank Close
NUCLEAR POWER Maxwell Irvine
NUCLEAR WEAPONS
  Joseph M. Siracusa
NUMBER THEORY Robin Wilson
NUMBERS Peter M. Higgins
NUTRITION David A. Bender
OBJECTIVITY Stephen Gaukroger
OCEANS Dorrik Stow
THE OLD TESTAMENT
  Michael D. Coogan
THE ORCHESTRA D. Kern Holoman
ORGANIC CHEMISTRY
  Graham Patrick
ORGANIZATIONS Mary Jo Hatch
ORGANIZED CRIME
  Georgios A. Antonopoulos and
  Georgios Papanicolaou
ORTHODOX CHRISTIANITY
  A. Edward Siecienski
PAGANISM Owen Davies
THE PALESTINIAN-ISRAELI
  CONFLICT Martin Bunton
PANDEMICS Christian W. McMillen
PARTICLE PHYSICS Frank Close
PAUL E. P. Sanders
PEACE Oliver P. Richmond
PENTECOSTALISM
  William K. Kay
PERCEPTION Brian Rogers
THE PERIODIC TABLE
  Eric R. Scerri
PHILOSOPHY Edward Craig

PHILOSOPHY IN THE ISLAMIC
  WORLD Peter Adamson
PHILOSOPHY OF BIOLOGY
  Samir Okasha
PHILOSOPHY OF LAW
  Raymond Wacks
PHILOSOPHY OF SCIENCE
  Samir Okasha
PHILOSOPHY OF RELIGION
  Tim Bayne
PHOTOGRAPHY Steve Edwards
PHYSICAL CHEMISTRY Peter Atkins
PHYSICS Sidney Perkowitz
PILGRIMAGE Ian Reader
PLAGUE Paul Slack
PLANETARY SYSTEMS
  Raymond T. Pierrehumbert
PLANETS David A. Rothery
PLANTS Timothy Walker
PLATE TECTONICS Peter Molnar
PLATO Julia Annas
POETRY Bernard O'Donoghue
POLITICAL PHILOSOPHY
  David Miller
POLITICS Kenneth Minogue
POPULISM Cas Mudde and
  Cristóbal Rovira Kaltwasser
POSTCOLONIALISM Robert Young
POSTMODERNISM Christopher Butler
POSTSTRUCTURALISM
  Catherine Belsey
POVERTY Philip N. Jefferson
PREHISTORY Chris Gosden
PRESOCRATIC PHILOSOPHY
  Catherine Osborne
PRIVACY Raymond Wacks
PROBABILITY John Haigh
PROGRESSIVISM Walter Nugent
PROHIBITION W. J. Rorabaugh
PROJECTS Andrew Davies
PROTESTANTISM Mark A. Noll
PSYCHIATRY Tom Burns
PSYCHOANALYSIS Daniel Pick
PSYCHOLOGY Gillian Butler and
  Freda McManus
PSYCHOLOGY OF MUSIC
  Elizabeth Hellmuth Margulis
PSYCHOPATHY Essi Viding
PSYCHOTHERAPY Tom Burns and
  Eva Burns-Lundgren

PUBLIC ADMINISTRATION
    Stella Z. Theodoulou and Ravi K. Roy
PUBLIC HEALTH Virginia Berridge
PURITANISM Francis J. Bremer
THE QUAKERS Pink Dandelion
QUANTUM THEORY
    John Polkinghorne
RACISM Ali Rattansi
RADIOACTIVITY Claudio Tuniz
RASTAFARI Ennis B. Edmonds
READING Belinda Jack
THE REAGAN REVOLUTION Gil Troy
REALITY Jan Westerhoff
RECONSTRUCTION Allen C. Guelzo
THE REFORMATION Peter Marshall
RELATIVITY Russell Stannard
RELIGION IN AMERICA Timothy Beal
THE RENAISSANCE Jerry Brotton
RENAISSANCE ART
    Geraldine A. Johnson
RENEWABLE ENERGY Nick Jelley
REPTILES T. S. Kemp
REVOLUTIONS Jack A. Goldstone
RHETORIC Richard Toye
RISK Baruch Fischhoff and John Kadvany
RITUAL Barry Stephenson
RIVERS Nick Middleton
ROBOTICS Alan Winfield
ROCKS Jan Zalasiewicz
ROMAN BRITAIN Peter Salway
THE ROMAN EMPIRE
    Christopher Kelly
THE ROMAN REPUBLIC
    David M. Gwynn
ROMANTICISM Michael Ferber
ROUSSEAU Robert Wokler
RUSSELL A. C. Grayling
RUSSIAN HISTORY Geoffrey Hosking
RUSSIAN LITERATURE Catriona Kelly
THE RUSSIAN REVOLUTION
    S. A. Smith
SAINTS Simon Yarrow
SAVANNAS Peter A. Furley
SCEPTICISM Duncan Pritchard
SCHIZOPHRENIA Chris Frith and
    Eve Johnstone
SCHOPENHAUER
    Christopher Janaway
SCIENCE AND RELIGION
    Thomas Dixon

SCIENCE FICTION David Seed
THE SCIENTIFIC REVOLUTION
    Lawrence M. Principe
SCOTLAND Rab Houston
SECULARISM Andrew Copson
SEXUAL SELECTION Marlene Zuk and
    Leigh W. Simmons
SEXUALITY Véronique Mottier
SHAKESPEARE'S COMEDIES
    Bart van Es
SHAKESPEARE'S SONNETS AND
    POEMS Jonathan F. S. Post
SHAKESPEARE'S TRAGEDIES
    Stanley Wells
SIKHISM Eleanor Nesbitt
THE SILK ROAD James A. Millward
SLANG Jonathon Green
SLEEP Steven W. Lockley and
    Russell G. Foster
SOCIAL AND CULTURAL
    ANTHROPOLOGY
    John Monaghan and Peter Just
SOCIAL PSYCHOLOGY Richard J. Crisp
SOCIAL WORK Sally Holland and
    Jonathan Scourfield
SOCIALISM Michael Newman
SOCIOLINGUISTICS John Edwards
SOCIOLOGY Steve Bruce
SOCRATES C. C. W. Taylor
SOUND Mike Goldsmith
SOUTHEAST ASIA James R. Rush
THE SOVIET UNION Stephen Lovell
THE SPANISH CIVIL WAR
    Helen Graham
SPANISH LITERATURE Jo Labanyi
SPINOZA Roger Scruton
SPIRITUALITY Philip Sheldrake
SPORT Mike Cronin
STARS Andrew King
STATISTICS David J. Hand
STEM CELLS Jonathan Slack
STOICISM Brad Inwood
STRUCTURAL ENGINEERING
    David Blockley
STUART BRITAIN John Morrill
SUPERCONDUCTIVITY
    Stephen Blundell
SUPERSTITION Stuart Vyse
SYMMETRY Ian Stewart
SYNAESTHESIA Julia Simner

SYNTHETIC BIOLOGY Jamie A. Davies
SYSTEMS BIOLOGY Eberhard O. Voit
TAXATION Stephen Smith
TEETH Peter S. Ungar
TELESCOPES Geoff Cottrell
TERRORISM Charles Townshend
THEATRE Marvin Carlson
THEOLOGY David F. Ford
THINKING AND REASONING
    Jonathan St B. T. Evans
THOMAS AQUINAS Fergus Kerr
THOUGHT Tim Bayne
TIBETAN BUDDHISM
    Matthew T. Kapstein
TIDES David George Bowers and
    Emyr Martyn Roberts
TOCQUEVILLE Harvey C. Mansfield
TOPOLOGY Richard Earl
TRAGEDY Adrian Poole
TRANSLATION Matthew Reynolds
THE TREATY OF VERSAILLES
    Michael S. Neiberg
TRIGONOMETRY
    Glen Van Brummelen
THE TROJAN WAR Eric H. Cline
TRUST Katherine Hawley
THE TUDORS John Guy
TWENTIETH-CENTURY BRITAIN
    Kenneth O. Morgan
TYPOGRAPHY Paul Luna
THE UNITED NATIONS
    Jussi M. Hanhimäki
UNIVERSITIES AND COLLEGES
    David Palfreyman and Paul Temple

THE U.S. CONGRESS
    Donald A. Ritchie
THE U.S. CONSTITUTION
    David J. Bodenhamer
THE U.S. SUPREME COURT
    Linda Greenhouse
UTILITARIANISM
    Katarzyna de Lazari-Radek and
    Peter Singer
UTOPIANISM Lyman Tower Sargent
VETERINARY SCIENCE
    James Yeates
THE VIKINGS Julian D. Richards
VIRUSES Dorothy H. Crawford
VOLTAIRE Nicholas Cronk
WAR AND TECHNOLOGY
    Alex Roland
WATER John Finney
WAVES Mike Goldsmith
WEATHER Storm Dunlop
THE WELFARE STATE David Garland
WILLIAM SHAKESPEARE
    Stanley Wells
WITCHCRAFT Malcolm Gaskill
WITTGENSTEIN A. C. Grayling
WORK Stephen Fineman
WORLD MUSIC Philip Bohlman
THE WORLD TRADE
    ORGANIZATION Amrita Narlikar
WORLD WAR II Gerhard L. Weinberg
WRITING AND SCRIPT
    Andrew Robinson
ZIONISM Michael Stanislawski
ÉMILE ZOLA Brian Nelson

Available soon:

JANE AUSTEN Tom Keymer
JOHN STUART MILL Gregory Claeys
GEOMETRY Maciej Dunajski

COGNITIVE BEHAVIOURAL
    THERAPY Freda McManus
HUMAN RESOURCE
    MANAGEMENT Adrian Wilkinson

For more information visit our website

www.oup.com/vsi/

Raymond T. Pierrehumbert

PLANETARY SYSTEMS

A Very Short Introduction

OXFORD
UNIVERSITY PRESS

Great Clarendon Street, Oxford, OX2 6DP,
United Kingdom

Oxford University Press is a department of the University of Oxford.
It furthers the University's objective of excellence in research, scholarship,
and education by publishing worldwide. Oxford is a registered trade mark of
Oxford University Press in the UK and in certain other countries

Published in the United States of America by Oxford University Press
198 Madison Avenue, New York, NY 10016, United States of America

British Library Cataloguing in Publication Data
Data available

Library of Congress Control Number: 2021943452

ISBN 978-0-19-884112-8

Printed and bound by
CPI Group (UK) Ltd, Croydon, CR0 4YY

For my children and grandchildren.
May their curiosity and sense of wonder never
diminish.

Contents

List of illustrations xvii

1 Beginnings 1

2 Creation revealed 14

3 Beautiful theories, ugly facts 31

4 What are planets made of? 47

5 A grand tour of exoplanets 76

6 Planetary climate and habitability 96

7 How it all ends 122

References and further reading 131

Index 133

List of illustrations

1   The Pillars of Creation **12**
    NASA, ESA and the Hubble Heritage
    Team (STScI/AURA).

2   The electromagnetic
    spectrum **16**

3   The Planck function **18**

4   How disk and star properties
    determine spectral energy
    distribution **24**

5   The Atacama Large
    Millimeter/Submillimeter
    Array **26**
    ESO/C. Malin.

6   Images of three nearby
    protoplanetary disks **28**
    Created from https://almascience.
    eso.org/alma-data.

7   The structure of a
    protoplanetary disk **39**

8   Luminosity vs. time for main
    sequence stars **55**

9   Elemental composition of the
    Sun **57**

10 A few representative
    planetary structures **74**

11 The two main methods for
    detecting and characterizing
    exoplanets **78**

12 Scatter plot of planet size and
    planet mass vs. instellation **84**

13 Scatter plot of mass vs.
    radius **86**

14 Trappist 1e and 1f as seen
    in the night sky of Trappist
    1d **89**
    Adapted from Samuel Zeller, 'A shot
    of the Clock Tower and London at
    Night', photo via Good Free Photos,
    https://www.goodfreephotos.com.

15 Orbits of the Trappist 1
    system **93**

16 Determination of the
    radiating temperature
    of a planet by energy
    balance **98**

17 Sketch illustrating the
    operation of the greenhouse
    effect **104**

18 Sketch illustrating how
    energy balance determines
    the inner edge of the
    habitable zone **110**

19 Sketch illustrating how
    energy balance determines
    the outer edge of the habitable
    zone **115**

Planetary Systems

Chapter 1
Beginnings

Matter is gregarious. From the almost featureless *tabula rasa* of the primordial Universe emerging from the Big Bang 13.8 billion years ago, matter clumps under the inexorable tug of gravity to form organized structures on a variety of scales, all the way up to the scale of the whole Universe. One of these clumps became our home galaxy, the Milky Way. The Milky Way was among the very first galaxies formed, and is very nearly as old as the Universe itself. Something over four and a half billion years ago, in a spiral arm on the outskirts of the Milky Way, our Solar System came into being and began its long voyage through time and space.

Most of the Universe is made of a mysterious substance called 'dark matter' and an even more mysterious substance called 'dark energy'. Ordinary matter—the stuff our Solar System and ourselves are made of—makes up just 5% of the Universe, with dark matter accounting for 25% and the remainder being dark energy. Ordinary matter is called 'baryonic', after the heavy particles (e.g. protons and neutrons) of which it is mostly made. Shortly after the birth of the Universe in the Big Bang, about 75% of the baryonic matter was hydrogen, with almost all of the rest being helium. Things haven't changed much since, but the tiny bits of stardust which have accumulated contain the heavier elements that make it possible to form beings like ourselves, and the planets on which we live.

A protostar is born

In order for matter to collapse into structures, it has to be able to get rid of the heat generated by the compression of the gas. Compressed gas at galactic scales heats up for much the same reason that a bicycle tyre heats up when you pump it up rapidly. The air in the tyre heats up because of the energy you pump into it through the action of your muscles. Gas in a cloud collapsing under its own gravity heats up because of the energy imparted by molecules falling inwards through a gravitational field. Baryonic matter can get rid of heat through the glow of emitted microwave, infrared, visible, and ultraviolet (UV) radiation. Dark matter cannot. That is why it is called 'dark'. Nor does it interact strongly enough with baryonic matter to transfer its heat there. Dark matter can get rid of the heat of compression only through inefficient means such as evaporating some of its own substance to the intergalactic void, and this does allow it to collapse into a roughly spherical dark matter halo surrounding almost all galaxies, but there the process pretty much ends. Baryonic matter continues to condense, though, to form the familiar disk shape of galaxies.

Within these disks, many galaxies, including our own, form spiral arms which are moderately denser than their surroundings. And within these spiral arms, yet denser Giant Molecular Clouds form. Although denser than their surroundings, they are not very dense. A volume of a Giant Molecular Cloud the size of the Earth would have a mass of just 360 kilograms (kg), and if squashed down to a manageable size could be carried down the stairs by two strong movers. Still, Giant Molecular Clouds are dense enough that most of the hydrogen in them forms into hydrogen molecules (H_2) consisting of two hydrogen atoms. That is why the clouds are called 'molecular'. Within one such Giant Molecular Cloud, a smaller clump began to form, and the more matter there is in a given volume, the stronger its gravity, so the more matter is sucked in. This clump ultimately gave rise to our Solar System. The chemical composition of the Solar System suggests that the process got its initial nudge from the explosion of one or more nearby supernovae, which were themselves the product of massive, short-lived stars that formed in the same Giant Molecular Cloud as the Sun. In fact, hundreds to thousands of other stars, ranging from a few percent of the Sun's mass to upwards of ten times the Sun's mass formed in the same cloud.

From that point on, things happened very quickly, at least in the way astronomers reckon time. After just ten thousand years, a luminous concentration of matter formed at the centre of the collapsing cloud. At this stage, the heat of the embryonic Sun was generated by gravitational collapse. Such objects are called protostars. As the gas collapsed, it swirled into a vortex, for reasons similar to the formation of a swirling vortex when water goes down a bathtub drain. The swirl flattened the collapsing cloud of gas into a spinning disk. In the outer regions of such a disk, the centrifugal force of the spinning gas mostly balances the tug of gravity, and gas orbits in a ring without much falling inwards. Closer to the proto-Sun, though, friction slowed the spin and allowed gas (mostly hydrogen) to accrete onto the proto-Sun. The proto-Sun grew, and its increasing gravity pulled in yet more mass. This process happened so fast that half of the ultimate mass of the proto-Sun accreted in only 35,000 years. At first, the proto-Sun was shrouded in dust. The dust in question was composed of grains of various minerals, a variety of ices, and even some organic haze particles that can be synthesized abiotically from the gases surrounding the protostar. Sometime between a half million and a million years later, the proto-Sun became hot enough that its radiation blasted away the surrounding gas and dust of the Giant Molecular Cloud, and as the dust cleared the proto-Sun would have become visible as a point of light to any astronomers observing the Solar System from afar at the time. The flattened disk, however, had not yet dissipated. Today, astronomers can see numerous objects of this type in our galaxy. They are called T-Tauri stars, and are known by their wild variability and high output of X-rays. T-Tauri stars are often accompanied by pulsating jets of material ejected from the disk.

The disks that surround protostars and persist into the earlier parts of their T-Tauri stages are called protoplanetary disks, because that's where planets are born.

Formation of gas and ice giant planets

All this time, something interesting was going on in the outer reaches of the disk. The material of the disk was mostly hydrogen, with almost all of the rest (about a quarter of the total mass) made up by helium. But the key action was taking place in the remaining 1% or so which is composed of heavier elements—mainly carbon, nitrogen, oxygen, magnesium, silicon, sulphur and iron. We'll learn how this stardust is made in Chapter 4. Some of the iron vapour condensed out into grains of metallic iron. Carbon condensed into tiny granules of graphite, the crystalline form of carbon used to make pencils. Other atoms joined hands to form molecules—here, a lot of silicon got together with oxygen to make silicate—SiO_3, which combined with other atoms to make a palette of minerals. There, the silicate got together with a magnesium atom to make a mineral called enstatite, which, too, condenses into a dust grain. Oxygen got together with the abundant hydrogen to make water (H_2O). Out in this part of the disk, it was cold enough for water to form ice. The ice helped the small dust grains stick together to form dust-bunny-like clumps which aggregated to form pebbles. There are many ices, including methane (natural gas, CH_4), carbon dioxide (CO_2), ammonia (NH_3), and the nitrogen molecule N_2 which makes up most of the Earth's atmosphere. Many of these could only condense out in the further-flung, colder parts of the disk.

In a process akin to droplets of cloud water aggregating through collisions until the droplets are massive enough to form rain, the dust grains kept aggregating through pebble size and continued onwards to the point where the clumps had enough gravity to compress the fluff into a hard kernel which could resist the tendency to shatter upon collisions. Such objects are a kilometre (km) or more in size, and are called planetesimals. The process is like the formation of a cosmic deluge, except that instead of raindrops we have ice/rock/iron assemblages which rapidly grow to the size of the Moon. The gas of the disk, which was still present, helped bring the small fluffy aggregates together and kept them from colliding so hard as to fragment, allowing them to grow inexorably in mass. When one of these cores grew to around 10 Earth masses, it triggered a process of runaway hydrogen accretion: the gravity sucked in hydrogen from the disk, which was compressed by the gravity of the core, adding yet more mass and hence yet more gravity, further accelerating the accretion. It is a true runaway, in which all the hydrogen available to the core was sucked in after just a finite amount of time. This was the process that gave rise to the hydrogen-dominated gas giant planets Jupiter and Saturn. By cosmic standards, the whole process was over in the blink of an eye. Core growth to the runaway threshold took a mere 500,000 years, a fifth of the span of time since Genus Homo first walked on Earth. The gas giants grew to their full size after just eight million years. For gas giants to exist at all, they need to form quickly, before the disk is cleared of gas by the radiation and stellar wind from the brightening protostar.

Even after the hydrogen was depleted, there was enough ice left over to form some quite large planets. In our Solar System, Uranus and Neptune are called 'ice giants' because they have a density that is compatible with being built mainly from various ices, including water ice. Their density is too high for them to be made mainly of hydrogen and helium, as are the gas giants. In the hot, high-pressure interior, these ices would no longer take the form of the conventional low-pressure ice such as one might skate on, but would transform into a variety of exotic, high-density fluids and semi-solids. The true composition of Uranus and Neptune, and the proportion of rocky material they contain, is at present still subject to considerable uncertainty.

After forming Uranus and Neptune and a smattering of icy moons, there was still *a lot* of ice left, made of a variety of different ices and mostly fairly dusty. Much of it went into making up the thousands of iceballs of various sizes out beyond the orbit of Pluto, known as the Kuiper Belt. Many people were distressed when the Pluto of their youth was demoted from the status of a planet to that of a Kuiper Belt object, but to look on the bright side we haven't lost a planet, but gained a Kuiper Belt object—and a very interesting one it is, as revealed by the flyby of the New Horizons spacecraft which is now making its way deep into the Kuiper Belt. Yet further out, the Oort Cloud formed. The Oort Cloud is a cloud of comets loosely bound to the Sun by gravity. By some reckonings it extends nearly halfway out to our nearest stellar neighbour Proxima Centauri. It has never been directly observed, but its existence is inferred from the population of long-period comets that visit the inner Solar System. Probably, the Oort Cloud bodies were formed in the same vicinity as the outer planets, but were flung outwards by the gravitational action of the gas giants.

Evolution of the proto-Sun

Meanwhile, back at the proto-Sun, mass continued to accrete during the time the first planets were forming. The rapid contraction of the proto-Sun paused when the core temperature rose to a still-chilly 200 Kelvins (200K, which is 200 degrees above absolute zero, or equivalently −73 degrees Celsius (°C)), at which point the interior pressure had built up sufficiently to counteract the force of gravity. As the outer shell of the proto-Sun continued to accrete mass, the interior continued to heat up, and when it reached 2,000K (1,727°C), the H_2 began to dissociate into its constituent atoms. Because energy then went into breaking up H_2 rather than increasing temperature, the proto-Sun entered a second stage of rapid contraction, and the power output of the proto-Sun increased. The rapid contraction again halted by the time the core temperature reached about 20,000K, at which point the dissociation of H_2 was essentially complete and the material of the proto-Sun became less compressible.

When the proto-Sun became visible as a T-Tauri star, it was still accreting mass from the disk. All this time, a portion of the outer disk was exposed to the proto-sunlight, causing it to evaporate away into space. Eventually, the mass loss eliminated the ability of the outer disk to feed the inner disk at a rate sufficient to sustain it against mass lost by accretion onto the proto-Sun. Thereafter, gas and dust rapidly drained from the inner disk, opening a hole roughly the size of the Earth's current orbit around the Sun. Then, the inner edge of the remaining disk was directly exposed to proto-sunlight, and the intense heating dissipated its gas and dust in short order. It is fortunate that not all the mass of the disk drained away into the proto-Sun, otherwise there would be nothing left to make the rest of the Solar System. As it was, the amount of disk material that wound up in planets made up only a few percent of the total mass of the system. It was only a few percent, but it is why we are here to be able to think about such things.

The youthful exuberance of the T-Tauri stage was thus spent after about ten million years. The proto-Sun at that point had attained its full mass, but it still had a long way to go before it became a fully-fledged star. The light emitted from the young Sun at this point drew its energy from the gravitational energy associated with its continued slow contraction. The young Sun put out more power at this stage than it does at present because it was considerably bigger than its ultimate size, but had a similar surface temperature to today's Sun. The contraction is in fact driven by the energy lost from the surface in the form of emitted light. The gravitational effect of cramming the mass of the young Sun into a smaller volume had the effect of steadily increasing the pressure and temperature in its core, which is perhaps counter-intuitive given that the object is losing energy and thus might be expected to cool down. The fusion of hydrogen into helium which powers most fully-fledged stars (see Box 1) requires a core temperature of around 20 million Kelvin. For stars of the mass of our Sun, that takes a hundred million years to reach. By the time fusion was triggered in the Sun, the gas and dust of the whole disk had been long dissipated, and the planets had reached very nearly their mature masses.

Box 1 What makes the Sun shine?

Before conservation of energy emerged as a foundational principle of science, there was little reason to be concerned about the question of what kept the Sun shining and the Earth warm. By the mid-1800s, though, the idea had taken hold that the Sun's energy had to come from *somewhere*. Coal was king at the time, but it was quickly ascertained that chemical energy, such as the burning of coal, could keep the Sun alight for a mere 5,000 years. In 1854 Hermann Helmholtz proposed that the energy of the Sun is supplied by the heat generated through its shrinkage under gravity—just as a cannonball falling through the Earth's gravity gains energy, molecules of the Sun gain energy, which can be converted to heat, by getting closer to the Sun's centre as the Sun shrinks. The rate of shrinkage needed is only 80 metres (m) per year, so did not contradict any observations. Helmholtz concluded that the Sun could have been shining for 21 million years, which at the time seemed a quite adequately long period, as there was not yet any real idea of how old the Earth is. Given that it is the Sun that keeps the Earth from being a lifeless frozen ball of rock and ice, the great Charles Darwin was rather uncomfortable with ascribing such a young age to the Sun. Geologists, especially Charles Lyell, also found a young Sun hard to reconcile with the sedimentary record. However, emerging (though ultimately incorrect) estimates of the age of the Earth based on its cooling rate seemed to support the idea of a fairly young Earth. The discomfort continued for some time, accompanied by various creative (but swiftly disproven) ideas for extending the age of the Sun, but relief did not come until the first half of the 20th century. The evolution of the new science of radioactivity allowed rocks to be dated with some precision, and by 1927 Arthur Holmes had set the age of the Earth at billions of years. About the same time, Einstein's famous mass-energy relation $E = mc^2$, together with the realization that an atom of helium had less mass than the sum of its parts, led Sir Arthur Eddington to realize in 1920 that the fusion of hydrogen into helium could release massive amounts of energy, enough to power the stars for tens or hundreds of billions of years. The idea remained speculative, though, until 1938, when Hans Bethe completed the formidable task of computing the rate at which fusion could release energy.

Formation of inner rocky planets

Now we flash back to the story of the inner rocky planets—Mercury, Venus, Earth, and Mars. In the inner regions of the disk, rocky cores did not grow rapidly enough to trigger runaway hydrogen accretion, and so gas giants did not form. Ice giants couldn't form because it was too hot for ice to survive for long. However, formation of rock and iron cores could proceed in much the same way as it did in the outer disk. The specifics of how the inner rocky Solar System planets formed is still a matter of considerable controversy, but it is generally thought that rocky core growth proceeded more slowly than in the outer disk, at least in part because the absence of ice makes it harder for clusters of dust grains to grow. As in the outer disk, growth of planetesimals relied on the drag against particles provided by gas. It took a million years or so for planetesimals to grow to the Mars-size objects called planetary embryos. By this time the cores of the gas giant planets were already fully formed. The gas, and most of the dust, of the disk was gone by the time the disk was ten million years old, and the growth of rocky planetary embryos had to be complete by this time. When the gas and dust of the inner disk drained away into the protostar, it left the rocky embryos behind. Another hundred million years had to pass before infrequent giant collisions brought Earth and Venus up to their present sizes. One such giant impact, most probably the last one, gave rise to our Moon.

Planetary systems everywhere

A Giant Molecular Cloud leaves behind a star cluster after radiation from the hot young stars blows away the remnant gas. If stars in the cluster are tightly bound gravitationally, they form a globular cluster which sticks together during its voyage around the galaxy. A globular cluster is not a good place to be if you are a habitable planet, since the star density is high and eventual close encounters of the worst kind disrupt planetary orbits and fling planets out into the interstellar void. Open clusters are weakly bound, and their stars eventually leave home, sometimes in small caravans known as moving groups. Our own Solar System came from such an open cluster, although if we were once part of a moving group our companions seem to have long since gone their own way.

This is how our Solar System was born. The same process is going on right now, with some variations on the theme according to the mass of the star being formed. It is happening at a rate of about one Solar mass worth of stars a day within around a thousand Giant Molecular Clouds in the Milky Way. On a clear night, you can see the nearest stellar nursery as a nebula in the constellation Orion. In the heart of the Orion Nebula lies the Trapezium cluster—an open cluster of stars much like the birthplace of the Solar System. The most readily identifiable part of the Trapezium consists of four bright young stars clustered within a bit over a light year of each other, with masses fifteen to thirty times the mass of the Sun. Like the Trapezium the much-loved constellation Pleiades, or Seven Sisters, is an open cluster of massive bright stars that formed within the past hundred million years. The Giant Molecular Cloud in which the Pleiades were born has already dissipated. The stars in the Trapezium and Pleiades won't last long, so enjoy them while you can.

In the opening sequence of the movie *Contact*, based on a screenplay by Carl Sagan and Ann Druyan, the viewer follows a front of radio transmissions as it expands outwards from Earth. Five thousand light years out, the wave passes through the Eagle Nebula, a Giant Molecular Cloud similar to the one in which our Solar System was born. Here, one is treated to an image—a real image, not some artist's fantasy—of the spectacular star-forming region known as the Pillars of Creation (Figure 1). Later, when Ellie Arroway travels 25 light years out to Vega in The Machine, we see a rather good artist's rendition of a protoplanetary disk around the star. Ellie exclaims, 'It's beautiful!'—as indeed it is. Vega really does sport a debris disk, but it is far more tenuous than that depicted in the film, and not a site of active planet formation. The actual debris disk has a mass somewhat less than our Moon, and if you were at Vega you'd hardly notice it was there unless you brought along a really good infrared telescope. In Chapter 2, we'll get to see actual protoplanetary disks in action.

Each stellar nursery, in its lifetime, will give birth to some hundreds to thousands of stars, most of which will have planets. Some will be high mass stars which burn brightly but briefly, snuffing out any life on their planets before it barely has a chance to get underway. They end their lives inflating to red giants or by exploding as supernovae, variously leaving behind white dwarf stars, neutron stars, or black holes. Others will be very low mass stars, with a fraction of the mass of the Sun, which will still be shining a trillion years from now when the Universe is dark and spread thin and no more stars are being formed. And in the

1. **The Pillars of Creation, a star-forming region in the Eagle Nebula.**

middle, are stars like the Sun, which live for around ten billion years before turning into a red giant and engulfing the inner planets. We are now in the Sun's middle age, about halfway to Earth's ultimate habitability crisis. The Milky Way is not especially out of the ordinary as spiral arm galaxies go. Stars and their planets are being born in similar stellar nurseries in any of the roughly fifty to a hundred billion spiral galaxies in the Universe. In fact, it is the bright young short-lived and newly formed stars in the spiral arms that primarily makes them visible. Stars and planets can also form in galaxies without spiral arms (e.g. elliptical galaxies), although they do so at a slower rate.

Until relatively recently, there was no real need for a term referring in general to the kind of object our Solar System is. It was the only known object of its type. We knew of stars but no planets outside the Solar System. We had no ability to observe planet formation in action. That has all changed, but so recently that there is no generally agreed term in the astronomical community for a star and all the gravitationally bound objects surrounding it. The term 'planetary system' has begun to gain currency to describe such objects, and it is the term we adopt to refer to a star and all the bodies gravitationally bound to it—the planets whether rocky, gassy, or icy, their moons, the asteroids, comets, and the far-flung icy bodies that make up Kuiper Belts. Our own planetary system contains only one star, but other planetary systems commonly contain two or even three stars. While the same general processes that formed our Solar System were also operating in the formation of other planetary systems, the end result of the process can yield planetary systems very unlike our own. Now that the Solar System isn't the only example of a planetary system subject to study, and now that we can in effect peer back in time and observe processes such as those that occurred billions of years ago when our Solar System was being born, we can begin to appreciate how our home planetary system, and indeed our home world, is or isn't special. The veil has been lifted, and this book provides a glimpse of what has been revealed.

Chapter 2
Creation revealed

The previous chapter outlined how we came to be, but how do we know it's not just another creation myth? This is the story we tell here, and it is no less awe-inspiring and humbling for being told in the language of science. The progress of science involves a constant interplay of observation and theory, with frequent titanic clashes between the two. In this chapter, we'll take a look at the key observations.

Visible light astronomy

Astronomers—and by this term I mean anybody who has ever gazed at the night sky—at first could probe the universe only through the medium of visible light. Visible light consists of the rainbow of colours of light ranging from red through violet, and is of course a species-relative term. Most species that live on or near the surface of Earth have evolved to make heavy use of the kind of light our Sun puts out most abundantly, and this is what defines 'visibility', although some species can see 'colours' somewhat outside the range of what we can see. They see the ultraviolets that are more violet than violet, and the infrareds that are deeper red than the deepest red we can see. Quite significant strides were made based on naked-eye astronomy, going back at least to the Babylonians and early Chinese observers. In the early 1600s the invention of the telescope, quickly adopted by Galileo in his astronomical studies, allowed us to observe the Universe in much greater detail, although still only for visible light.

Visualizing the invisible

In 1800, Frederick William Herschel discovered that sunlight split into colours by a prism contains a band beyond the red which was invisible to the human eye but was capable of powerfully heating a thermometer. With the discovery of these 'heat rays', the seeds of infrared astronomy were planted, although they did not blossom until the early 20th century saw the invention of more sensitive detectors. Meanwhile, throughout the course of the 19th century, one of the grandest unifications in physics was accomplished. Through the dedicated work of a handful of brilliant physicists, it was discovered that the forces of electricity and magnetism, known since ancient times, were in fact different aspects of the same phenomenon, dubbed electromagnetism. What's more, James Clerk Maxwell, who completed the formulation of the equations of electromagnetism that bear his name, found that the equations predicted waves which all moved through a vacuum at the same speed—what we would now call the speed of light—regardless of their wavelength. Radio waves (also discovered in the 19th century), infrared, visible light, ultraviolet (discovered in 1801 by Johann Wilhelm Ritter), and X-rays were all instances of electromagnetic radiation, distinguished by their wavelength. Visible light has wavelengths ranging from red light at .75 microns (millionths of a metre) to violet at .38 microns. A prism splits these colours up into the rainbow colours of the spectrum, and by analogy the broader range of electromagnetic waves, including those we cannot see with our unaided eyes, is called the electromagnetic spectrum.

An electromagnetic wave consists of two force fields—an electric field and a magnetic field—that move in concert. In a vacuum, all electromagnetic waves move at the same speed, denoted by the symbol c (for 'celerity'). This speed is generally referred to as 'the speed of light', although it is common to all forms of electromagnetic radiation and not just visible light. The electric field exerts a force on charged particles, while the magnetic field can cause a force on magnetized objects such as a compass needle. Time-varying magnetic fields also induce a current in a conducting wire, which is how both electric generators and radio receivers work. The different types of electromagnetic waves are distinguished by their wavelengths, as indicated in Figure 2. (1 metre = 1,000 millimetres = 1,000,000 microns; 1 millimetre = 1,000 microns). The wavelength is the distance over which the electric or magnetic field repeats itself. Waves can equivalently be distinguished by their frequency, defined as the number of peak-to-peak cycles per second seen by a stationary observer as the wave moves past. Since all electromagnetic waves move at the same speed, shorter wavelength waves have higher frequency, in inverse proportion to their wavelength. The unit 'cycles per second' also goes by the name 'Hertz' (Hz for short), in honour of Heinrich Hertz who was one of the great pioneers of electromagnetic radiation.

Infrared waves have longer wavelength than visible waves, microwaves have longer wavelength than infrared, and radio waves have longer wavelength still, extending into the hundreds of kilometres. Submillimetre waves could be considered to be either long wavelength infrared or short wavelength microwave, but

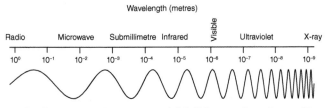

2. The electromagnetic spectrum. Visible light occupies only a small sliver of the spectrum from .75 $\times 10^{-6}$ metres (red) to .38 $\times 10^{-6}$ metres (violet) The wavelength in the sketch is not drawn to scale.

because of their importance in astronomy they have been given a name of their own. Monochromatic radiation consists of a single wavelength, but most natural sources of radiation produce a superposition of many different wavelengths, just as white light is a roughly equal mix of all the wavelengths of visible light. Modern astronomy exploits the entire electromagnetic spectrum, from long radio waves to short X-rays and even shorter gamma rays. Our main window into the formation of planetary systems lies in the infrared through microwave range.

Blackbody radiation

To understand how infrared and microwave astronomy have transformed our ability to observe the formation of planetary systems, it is necessary to know something about the way electromagnetic radiation is absorbed and emitted by matter. For matter, temperature is a measure of the average amount of energy stored in the motion of each of the molecules that makes up the object. When electromagnetic radiation interacts with matter, it acquires the temperature of the matter. The resulting collection of electromagnetic waves is called 'blackbody radiation'. The term 'black' arises from the fact that ideal blackbody radiation requires very strong interaction between the radiation and matter, one consequence of which is that any incident electromagnetic radiation is perfectly absorbed. Ideal blackbodies are, in this sense, indeed 'black'. Blackbody radiation is a blend of electromagnetic waves of various wavelengths, and the curve that characterizes the amount of energy in each wavelength going into a blend is called the spectral energy distribution. It is a remarkable property of blackbody radiation that the spectral energy distribution has a universal shape, now known as the Planck function (shown in Figure 3), that depends only on the temperature of the radiation. The wavelength at which the Planck function has its peak decreases in inverse proportion to temperature measured in degrees above absolute zero. The hotter the body, the shorter the wavelength at which it mainly emits. This property is called the

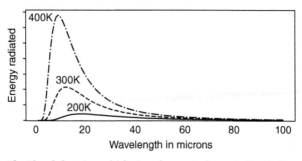

3. The Planck function, which gives the spectral energy distribution of blackbody radiation. For each wavelength, the spectral energy distribution gives the relative amount of energy radiated per unit time, from each square metre of the blackbody's surface. The temperature of the emitting body is labelled on each curve. The function is named after Max Planck, who provided the first theoretical explanation of the shape of the curve.

Wien–Boltzmann Displacement Law. The measure of temperature generally used in physics is Kelvins. A Kelvin degree is the same size as a familiar Celsius (Centigrade) degree, but the zero of the temperature scale is placed at absolute zero ($-273.15°C$) rather than the freezing point of water. A body with temperature 2,900K emits dominantly at a wavelength of 1 micron, just a bit longer than visible red light. For a body a tenth as hot, or 290K, the radiation would peak at 10 microns, deep in the infrared. This decrease of wavelength with temperature of the radiation lies behind the familiar notions of 'red hot' bodies, although in fact a body only needs to reach a temperature of 800K to radiate enough visible red light to appear red even though its emission peaks at 3.6 microns. The Wien–Boltzmann law is the reason that the colour put out by the light-bulb you buy at the shop is usually described in terms of a temperature. An old-style incandescent light has a colour temperature of around 2,800K, which is roughly the temperature of the tungsten filament inside when an electric current is flowing through it. Daylight has a colour temperature of 5,800K, which is the temperature of the surface layer of the Sun—called the photosphere—from which light escapes.

For any given wavelength, the Planck function increases with temperature—a hotter body emits radiant energy at a greater rate than a colder one. The total amount of energy emitted, summed up over all wavelengths, increases as the fourth power of temperature above absolute zero. This is called the Stefan–Boltzmann Law, and the proportionality constant is called the Stefan–Boltzmann constant, denoted by the symbol σ. The Planck function actually gives the rate at which energy is emitted from each square metre of the surface of an emitting body (such as a star, planet, or disk). Thus, for a body of a given temperature, the rate of energy emission increases in proportion to the surface area of the object. The energy emission rate—also known as power—is usually measured in units of Watts. An electric kettle consumes energy at a rate of about 1,000 Watts, and using this energy can bring a litre of water to the boiling point in six minutes. If T is the temperature of an ideal blackbody in Kelvins and A is its surface area in square metres, the power output P in Watts is

$$P = \sigma \times A \times T^4 \tag{2.1}$$

with $\sigma = 5.67 \times 10^{-8}$. A sphere of typical solid matter with a radius of 1 m and temperature of 290K (17°C, a typical room temperature) emits at a rate of 5,040 Watts. Most of this is in the infrared, with power output peaking at 10 microns. The power output of old-style incandescent light-bulbs is stamped on the end of the bulb, with 100 Watts putting out enough light to illuminate a fair sized room (although most of that power output is in the form of heat rather than visible light, given the 2,000–3,000K temperature of the tungsten filament that creates the light). If the Sun had a wattage stamped on it, the number would be 4 followed by 26 zeroes (4×10^{26}).

Dense solid or liquid materials, like dust particles or chunks of ice, radiate pretty much like an ideal blackbody. However, tenuous blobs of matter, such as gases, radiate much less effectively than an ideal blackbody. Chunks of matter that are good absorbers at a given wavelength turn out also to be good emitters, and vice versa.

This property was discovered by Gustav Kirchhoff in the mid-1800s in the course of his pioneering work on blackbody radiation. It is in fact why blackbodies are called 'black'—because they emit efficiently, they are also nearly perfect absorbers of radiation incident on them. This is why it is not absurd to say 'The Sun is a blackbody', although it appears anything but black.

Although it was known since the mid-1800s that blackbody radiation involved an exchange of energy between matter and electromagnetic radiation, it was not until the quantum revolution of the early 20th century that the process could be fully understood. In fact, it was study of blackbody radiation that provided much of the impetus for development of quantum theory. A fundamental principle of thermodynamics is that each degree of freedom that can store energy gets an equal share of the energy available to the whole system. Attempts to understand the equilibrium between electromagnetic waves and matter on this basis failed spectacularly. There are just too many ways to stuff energy into short wavelengths, leading to the absurd prediction that all matter would cool down infinitely quickly, emitting a burst of ultraviolet, X-rays, and even shorter waves in the process. It turned out that, in reality, although Maxwell firmly established electromagnetic radiation as a wave phenomenon, the old competing particle theory of light was not dead, but waiting out its time to be re-born. The quantum hypothesis introduced by Max Planck stipulated that electromagnetic radiation with a wavelength λ could not be emitted in just any arbitrary amount, but in fact could only be emitted or absorbed in integer units of a 'quantum' of energy, which is inversely proportional to λ, much like physical currency in the European Union is emitted in integer units of one cent. The quanta of electromagnetic energy are carried by particles called photons, and it takes a great deal of energy to emit even a single photon of a very short wavelength. This property gives rise to the fall-off in energy past the peak of the Planck spectrum. It is an unfathomable aspect of the quantum weirdness of the Universe that electromagnetic radiation (and most other things, including electrons) can simultaneously act as a wave and a particle, while really being neither. The quantum of energy is tiny, so discrete jumps in energy state are not evident in the motion of the objects of everyday life. Nonetheless, the underlying quantum nature of the Universe is manifest to our own senses in the fact that the Sun shines and that we are here to ponder such things.

Gases do absorb and emit electromagnetic radiation, but they do so very selectively, and the reason why is also tied in with the quantum nature of the Universe. The internal energy states of atoms and molecules are quantized, and can take on only a discrete set of values. Jumps between states are accompanied by emission or absorption of a photon. Since the differences in energy between any two states are then also quantized, a photon can only be absorbed or emitted with one of these allowable values, and by Planck's hypothesis this translates into a discrete set of wavelengths. Atoms and molecules are like little radio receivers and transmitters, tuned to broadcast or receive only a set of narrowly defined wavelengths. On a spectrograph these wavelengths show up as a set of thin bright lines for emitted radiation, or dark lines for radiation that has been absorbed by a gas it has traversed. These spectral lines give each molecule a unique fingerprint, which makes it possible to probe the chemical composition of distant stars and disks. Hydrogen interacts strongly with ultraviolet, because transitions in its electron states correspond to that wavelength range. Unless compressed to great density, hydrogen molecules are nearly transparent to infrared and submillimetre/microwaves, mainly because the symmetry of the molecule prevents its vibrations at these wavelengths from coupling efficiently to the electromagnetic field. Asymmetric molecules, such as carbon monoxide (CO), have a very orderly picket fence of spectral lines, making them valuable as a probe of the gas content of protoplanetary disks.

Infrared observations of young stellar objects

The canonical picture of an astronomer is of a crotchety old gent with a shock of white hair spending his lonely nights peering through the eyepiece of a telescope with the observatory dome open to the cold night air. This picture is quite outdated, and not just because a great many astronomers are now women. Nearly a century ago, photographic plates replaced the human eye as a means of making observations, and they have since been replaced by sophisticated digital imaging systems, like those in your smartphone camera but with much better resolution and sensitivity. What's more, a great deal of the work of modern astronomy does not involve taking images at all, but instead measuring spectral energy distributions. Lonely nights are still part of the package, but they are mostly spent babysitting finicky instrumentation or software. Although they are not images, spectral energy distributions contain a great deal of information about the geometry of the object being observed. This is most of the story of infrared observations of protoplanetary disks.

When a telescope is pointed at a young stellar object, it receives radiation both from the protostar and from the disk. The disk is heated by energy absorbed from the protostar and by the compression of gas falling into the disk from the surroundings, but because the disk has a very large area to cool from, it is considerably colder than the protostar. The inner regions of the disk, being more strongly illuminated by the protostar, are hotter than the outer regions. Although almost all of the mass of the disk is in the form of molecular hydrogen, the gas does not radiate very effectively. Most of the radiation emitted by the disk is emitted by the small amount of embedded dust, which radiates effectively as a blackbody. The dust shares the temperature of the gas in the disk, however, so it can be used to take the temperature of the disk.

Temperature controls wavelength, and for a given temperature the surface area of the disk controls the power emitted.

Taken together, the two properties make it possible to constrain the size and temperature distribution of protoplanetary disks from observations of their spectral energy distribution. The general idea of how this works is illustrated in Figure 4, for an idealized case in which the protostar has a temperature of 3,500K and a radius twice the radius of the Sun, and the disk has a temperature of 100K and a radius of 10 astronomical units. (An astronomical unit, or au, is the mean distance of the Earth from the Sun, or 150 million km.) Being colder, the disk radiates more weakly from each square metre of its surface, but this is made up for by the fact that the disk is much larger than the star and therefore has a much greater surface area. The disk radiates at longer wavelengths than the star, though, and this allows astronomers to distinguish the light from the young stellar object from what would be produced by a brighter protostar that had no disk. A star or protostar without a disk would emit very little infrared at wavelengths longer than 5 microns. The infrared excess shown in Figure 4 provides us with the fingerprint of the disk.

A more realistic disk would emit a blend of energy from a small hot inner disk and a larger but colder outer disk, plus everything in between. A hot inner disk would partly fill in the valley in emission between the protostellar emission and that of the outer disk. The farthest reaches of the disk are not detectable in the infrared, because they are too cold to emit much, even taking into account the area effect. Astronomers must also contend with the fact that, as seen from a telescope in the Solar System, disks are rarely presented fully face-on. They usually appear viewed at an angle, or even edge-on, which affects the relative emission from the protostar vs. disk; viewed fully edge-on it is even possible for the protostellar emission to be obscured.

Infrared astronomy can be conducted from Earth's surface, but it is impeded by the strong absorption of infrared by water vapour and clouds in the Earth's atmosphere. The effect can be reduced by siting observatories in high, cold, dry places like the Atacama

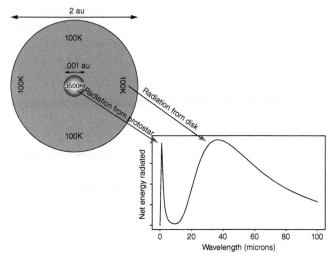

4. How disk and star properties determine the spectral energy distribution of the young stellar object. The example assumes the disk is viewed face-on. The disk, which is flat, has a uniform temperature of 100K, and the star, which is spherical, has a surface temperature of 3,500K. The spectral energy distribution in the inset gives the net energy per unit time emitted by the system, taking into account the relative areas of the disk and star. The star is not drawn to scale relative to the disk; 1 au is the mean distance of the Earth from the Sun, or about 150 million kilometres.

Desert in Chile, but visibility in the longer infrared wavelengths that say most about disk structure is still very limited. Space telescopes have smaller mirrors than ground-based telescopes, but they do not suffer the obscuring effect of the atmosphere. For infrared astronomy, this has been transformative. The launch of the Infrared Astronomical Satellite (IRAS) space telescope in 1983 opened up the full sky to full-spectrum infrared astronomy, but it was the Spitzer space telescope, launched in 2003, that in a voyage of discovery lasting nearly two decades really opened the floodgates.

Spitzer surveys have mapped 90% of the star-forming regions within 1,600 light years, yielding infrared spectra of over 2,000 young stellar objects. It is because of Spitzer that we know young stars almost invariably are surrounded by dusty disks when they first become visible a mere half million years into their existence. These disks typically extend out to at least 10 au from the star, and exhibit a hot inner disk without any evidence of a hole cleared out near the star.

Perhaps the most important result of the Spitzer survey was its clear determination of the lifetime of protoplanetary disks. An astronomer cannot sit and watch an individual disk for millions of years, but by taking a census of a large number of T-Tauri protostars, one catches systems of all ages and can then compare disk properties between younger and older systems, and in particular look for the age range in which the infrared excess disappears. The age of a protostar can be determined by measuring its spin (older stars spin up as they contract) and by looking for lithium depletion (lithium is consumed by a minor fusion reaction in the older protostars before they get hot enough to ignite fusion of hydrogen into helium). It was found that disk lifetimes around protostars of twice the mass of the Sun or less range from about two million years to a firm upper limit of ten million years, with a tendency for the more massive protostars to have shorter-lived disks. A common feature of the life cycle of all disks is that they persist for a few million years but then dissipate rapidly in under a half million years, leaving behind planets, protoplanets, and rocky debris that has formed before the disk dissipates.

The problem with infrared as a means of probing planet formation is that dust is too opaque to infrared radiation. Infrared can't probe the very early stages of disk formation, before the protostar has cleared out its local neighbourhood and become visible as a T-Tauri star. Later, you can see the surface of the protoplanetary disk, but not very far inside it. It is like looking into a muddy pond: you can see the muddy water, but not the fish swimming in it. To see our planetary fish, we need something akin to Superman's X-ray vision, which can see through the murk. The answer, though, is not in the X-rays, but at the other side of the electromagnetic spectrum: submillimetre waves and microwaves.

ALMA tell us

The Chajnantor Plateau, high up in the Atacama Desert of Chile, is surely one of the most barren places on Earth. Perched there stand the sixty-eight dish antennas that make up the Atacama Large Millimeter/Submillimeter Array—ALMA (Figure 5). ALMA observes the universe in the wavelength range between .3mm and 3.6mm, dipping into the shortest wave end of the microwave range.

There's a lot to be said in favour of doing astronomy in the microwave and submillimetre spectrum. Not only do submillimetre/microwaves allow us to peer deeper into the disk,

5. **ALMA: The Atacama Large Millimeter/Submillimeter Array.**

but they also make the farther, colder reaches of the disk visible, since these regions radiate dominantly at long wavelengths. Submillimetre/microwaves are also much less affected by the Earth's atmosphere than infrared waves, but they can still be blocked by very moist air, rain, and thick clouds—hence ALMA's high, dry location. But what about imaging? The smallest feature that can be resolved by a telescope gets smaller in inverse proportion to the size of the mirror but gets larger in proportion to the wavelength at which the observation is made. That would seem to favour the shorter infrared wavelengths if one wants a sharp image. The closest protoplanetary disk to Earth is TW Hydrae, 175 light years away. To image it with 1 micron infrared with a rather fuzzy resolution of .5 au would require a mirror 27 m in diameter—twice the size of the biggest existing infrared/visible telescope—even given perfect seeing conditions. The Extremely Large Telescope, now under construction, will have a 39.5 m mirror and open up disk imaging in the infrared, at least for the closest disks. (One wonders what name they will come up with for the successor to this telescope.) Using a technique called interferometry, it is possible to gang up an array of telescopes to act as a single giant telescope. With present technology, interferometry just barely works in the infrared. However, in the submm/microwave region it allows truly enormous telescopes to be built. Such a one is ALMA. As compared to infrared, the advantage of ALMA is not so much that it allows higher resolution. Rather, the longer wavelength allows a much more extensive portion of the disk to be mapped, in terms of both distance from the protostar and depth within the disk, while the use of interferometry overcomes the resolution disadvantage that would ordinarily be incurred by working at longer wavelengths.

Newborn planets sail through a sea of dusty hydrogen gas, leaving behind wakes that redistribute the matter in the disk. According to simulations, a very large planetary core can clear both dust and gas from a ring of the disk centred on the core's orbit, but even a relatively small core can push away some of the gas, causing a local dip in gas pressure, driving dust to its edges. The DSHARP survey used ALMA to image twenty nearby disks in search of substructures characteristic of planetary formation, All of the disks in the survey are T-Tauri stars or the equivalent, for which mass is actively accreting on to the protostar. Mass accretion can be estimated through the energy released in the maelstrom caused when new material falls onto the star.

Three examples from the DSHARP survey are shown in Figure 6. The bright bands are indicative of dust-rich regions, while the dark bands are regions that have been cleared of dust. HD163296 is visible as a bright blue star 326 light years distant in the constellation Sagittarius. The protostar which anchors the disk has

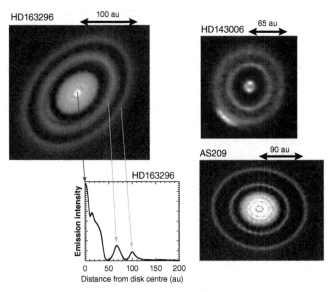

6. Images of three nearby protoplanetary disks, captured by the ALMA telescope array. Bright bands correspond to regions of strong emission of submillimetre radiation, and indicate high dust concentration. The inset shows a graph of the average emission as a function of distance from the protostar for HD163296.

a mass of 2 Solar masses, and is accumulating mass at a rate of about 1 Solar mass per ten million years, though this is a quite old disk at ten million years age, and the star will not accumulate much more mass before the disk dissipates. Systems anchored by massive hot blue stars like this are called Herbig Ae stars rather than T-Tauri stars, because of their characteristic spectra, but the systems do not differ in their essentials from T-Tauri systems. HD163296 exhibits a large well-defined disk with dust extending out to 100 au, with dark bands at 50 au and 90 au. Although dust has been cleared from these bands, it is known that gas has not been cleared, because a gas gap would interrupt the flow of mass through the disk that feeds accretion onto the protostar; this is the case for all the disks in the DSHARP survey. In some cases, the continued presence of gas in the gaps can be directly inferred from observations of CO lines in the ALMA images, though there is generally some indication of CO depletion indicative of gas accretion onto a protoplanet. There may be other explanations for the gaps, but at present it appears that by far the most likely explanation is that these are the signature of a gas or ice giant planet in the process of active formation.

HD143006 and AS209 are conventional T-Tauri stars with stellar masses of 1.7 and .83 Solar masses respectively. HD143006 is in the Upper Scorpius star-forming region, and is four million years old. The lower mass system AS209 is in the Ophiuchus N3a star-forming region, and is only a million years old. Both of these disks also exhibit the dark bands characteristic of planet formation, and AS209 shows some intriguing hints of disk substructure in the inner disk as well. Clear bands are ubiquitous throughout the DSHARP survey, and indeed are found in most ALMA disk images taken so far, though some show interesting additional structures such as spiral waves.

The ability to image the detailed structure of protoplanetary disks all the way to their outer reaches is ALMA's most dramatic contribution, but ALMA has vastly improved our understanding of disks in a number of other ways. By being able to penetrate deeper into disks, and discriminate large from small particles, ALMA has enabled accurate estimates of the dust mass in disks; by using estimates of the ratio of gas to dust, this provides estimates of total disk mass, giving an idea of the amount of mass available to form planets. Although ALMA can't detect hydrogen directly, it can detect other gases, such as CO, and through estimates of the ratio of carbon to hydrogen in interstellar gas provide an additional cross-check on disk mass. The shape of spectral lines of CO depends on temperature, providing a direct measurement of gas temperature. The Doppler shift in the wavelength of the CO lines (analogous to the change in pitch when an ambulance siren is moving towards or away from you) allows an estimate of the velocity of gas in the disk.

Building on earlier, more limited submillimetre/microwave observations, ALMA tells us that disk mass scales with stellar mass except for the few most massive stars. A typical disk mass is .3% to 3% of the mass of the star, with a median value of 1%. Since only 1% of disk mass is dust, there is really not much mass available to make rock/iron cores or rocky planets. That is why we find gas or ice giants, but not rock giants.

Chapter 3
Beautiful theories, ugly facts

The biologist Thomas Huxley said that the great tragedy of science is that one often sees a beautiful theory slain by an ugly fact. Ugly facts are much treasured in science, because they show the way to making a theory more beautiful. But without theory, there would be no basis for even saying which facts are ugly and which are beautiful.

The nebular hypothesis

Early thought about the origins of the Solar System had little to go on other than the observation that the Solar System consists of a massive Sun, plus a number of far less massive planets which orbit it all in the same direction, and nearly in a flat plane. In 1734, the Swedish polymath Emmanuel Swedenborg proposed that the Sun and all the planets condensed out of the same ball of gas, in what is probably the earliest statement of the nebular hypothesis. It is hard to categorize Swedenborg. He was a wide-ranging thinker, having delved into physiology of the brain as well as cosmology, but his methods were anything but mathematical and rigorous, and often tinged with mysticism. He may have had some external help in formulating the nebular hypothesis, as he claimed to have had regular conversations with beings on all the other planets in the Solar System, as well as in other planetary systems. These conversations extended even to God himself, and led him to found the religion which bears his name and still has many adherents today. One tenet of Swedenborgianism is that the Last Judgement already took place in 1757, but news hasn't reached the material world yet. Certainly one cannot fault him for lack of imagination. It is easy to laugh, but the progress of science proceeds through leaps of imagination. It's where new ideas come from. Swedenborg's ideas entered the scientific world through the efforts of the philosopher Immanuel Kant, who in 1755 published a more comprehensible form of the nebular hypothesis, stripped of Swedenborg's arcane metaphysical overlay. It entered something close to its modern form in the hands of the French mathematician Pierre-Simon Laplace, who in 1796 made the clear connection to Newtonian gravity. He inferred that the contracting blob of gas would swirl and flatten as it contracted, that much of the gas would go into making up the star, and the leftovers would make the planets.

Free fall time

With some help from Mr Newton, let's estimate the time it takes for the collapse of a large, tenuous region of gas to get underway. It is a property of Newtonian gravity that the gravitational force acting on a particle sitting on the surface of a sphere of matter is precisely the same as if all the matter in the sphere were concentrated in a point at its centre. Since this is true for every particle in the shell of gas and dust making up the outer edge of the sphere, the time to collapse to a point (if no other forces intervene) is the time it would take for a particle to free-fall to the centre of a sphere under the gravitational influence of a point mass. This time is finite, and can be calculated with pencil and paper from Newton's laws.

At typical densities of Giant Molecular Clouds, a ball of gas one light year in radius would contain roughly the mass of the Sun. Under gravity alone, it would take two million years to collapse to a point. The process starts slow, but accelerates rapidly. If no mass is shed by the time the sphere contracts to a tenth its original size, it will take only another 73,000 years to collapse to a point.

While the Universe as a whole is expanding, bits and pieces of it are collapsing like this all the time. It takes the enhanced density of a Giant Molecular Cloud to predispose a region to forming stars and their associated disks. Thankfully, there is a lot of randomness involved in the nudges that get an individual collapse started, so that we form stars with a whole range of masses from a tenth to a hundred times the mass of the Sun. (Recall that the mass of a protostar's surrounding disk is on the order of a percent of the mass of the protostar.) It would be a lonely galaxy if all the mass went into making just a few very massive stars, which burn out quickly, but we'll see in Chapter 4 that it would also be lonely if no massive stars were made at all.

The angular momentum problem

The gravitational collapse calculation is a good start, but it doesn't explain why there are planets, as opposed to all the matter being drained into the protostar. For that, we need to take a serious look at angular momentum. For a particle with mass M executing a circular motion with speed v at a distance r from the centre of rotation, the angular momentum is $M \times r \times v$. There must be some force, like gravity, or the tension in a rope, to keep the object moving in a circle instead of flying off in a straight line. For a collection of mass that is not subject to any external force, angular momentum is conserved as the system evolves. That is, within the system, angular momentum can be moved from one place to another, but it can't be created or destroyed, and the total angular momentum of the system remains always the same. In fact, conservation remains valid even under the action of a fairly broad range of forces, such as gravity, including most of the forces at play in protoplanetary disks.

To take an everyday example, consider an ice-skater spinning at a rate of once per 5 seconds, holding a pair of 5kg weights at arms' length, 1m apart straight out to either side. The speed of each weight is .63 metres per second (m/s), so the angular momentum of the pair is $5 \times .63 \times 1 = 3.15$kg $\cdot$ m^2/s. Now let's assume that this is a very, very skinny skater, so most of the angular momentum is in the weights, and the skater pulls the weights in to his body, so they are now 1/4m apart. The mass hasn't changed, but the distance to the centre of rotation has gone down by a factor of 4, so to conserve angular momentum the speed must go up by a factor of 4—to 2.52m/s, corresponding to a dizzying spin rate of once per .3 seconds. Because we haven't taken into account the angular momentum of the skater's body, this is an over-estimate of the increase in spin rate, but the general principle is the same as what actual figure skaters use.

If the object consists of many masses, then one simply sums up the angular momentum of the pieces. The angular momentum of a spinning sphere like the Sun can be calculated by splitting it up into a great many small bits and adding up.

An additional everyday example of the conservation of angular momentum is the formation of a bathtub vortex when the water is drained out. The water in the undrained bath contains a tiny, probably invisible amount of angular momentum. However, when the water is sucked down into the much smaller radius of the bathtub drain, it must spin wildly so as to conserve angular momentum. Something very similar happens in the course of collapse of a piece of a Giant Molecular Cloud, which also invariably has a small bit of angular momentum. As the cloud contracts under gravity, it must spin wildly to conserve angular momentum, and the resulting centrifugal forces flatten the collapsing sphere into a disk. As recognized by Laplace, this is why protoplanetary disks are disks, and it is why the Solar System (and as we now know, other planetary systems) lies pretty much in a plane.

Now, here comes an ugly fact, known as the 'angular momentum problem'. Almost all the mass of the Solar System is in the Sun. Only .14% of the mass of the Solar System is in the planets, mostly in Jupiter and Saturn. This is quite consistent with the observation that disk masses are typically only a percent or so of the protostar mass. One would then expect that most of the angular momentum of the Solar System is also in the Sun, insofar as angular momentum is concentrated in the Sun along with the mass it accretes. However, sunspot observations say that the Sun rotates about once per thirty days, on average. The rotation rate varies a bit from pole to equator, and modern measurements show the deep rotation rate is somewhat different from the surface. From this data, the angular momentum of the Sun is about 9.35×10^{41} angular momentum units. Now, Jupiter is much lower in mass, but it is orbiting at a distance over a thousand times the radius of the Sun, at a speed of 13,000m/s. This gives it a very high angular momentum of 1.9×10^{43} angular momentum units, or twenty times that of the Sun. Throw in the angular momentum of Saturn and the other planets, and we find that with under 1% of the mass of the Solar System, the planets contain 97% of its angular momentum. How can that be? This ugly fact was the main impediment to the acceptance of the nebular hypothesis for a century or more.

Angular momentum conservation is the reason that matter in the collapsing cloud does not all drain into the star, leaving nothing behind with which to form planets. But the mechanism works too well. As fluid spirals into the protoplanetary disk while conserving angular momentum, it will spin up to the point where centrifugal force balances gravity—at a distance called the centrifugal radius quite far out from the centre of the nebula. It would not even be possible for mass to accrete onto a protostar, and maybe stars should be doughnut-shaped rather than spheres. In order for mass to accrete onto a protostar, it has to have some way to shed its angular momentum to the outer regions of the disk, without also shedding its mass. These things sat for quite a while.

The angular momentum problem is somewhat less severe than originally thought, because it is now known that stars spin down as they age through interaction of the ejected stellar wind particles with the stellar magnetic field. That does not resolve the problem, though, since even young protostars are spinning so slowly that the material they accrete from needs to have lost a great deal of angular momentum. Some ideas about how this can be done began to emerge in the 1960s, through the work of the British astrophysicist Leon Mestel. It seems very likely that the resolution of the problem is to be found in friction of some sort. Under the action of gravity, the gas in the parts of the disk close to the protostar spins around faster than the more distant parts. Because the disk consists of gas and dust, rather than planets sailing through a hard vacuum, the fast parts of the disk rub up against the adjacent slower parts, slowing down the inner disk and spinning up the outer disk. The problem is that the friction caused by molecules rubbing up against each other is far too weak to transfer the amount of angular momentum needed. To get enough transport, the friction needs to be greatly enhanced by turbulence—the complex array of swirls and vortices of all sizes that one can see in a variety of everyday fluid phenomena, including water spilling over a weir. Turbulence requires some kind of instability to get it going. For example, if a layer of cold dense water is placed over hot, light water, the unstable density gradient will cause the dense water to mix downwards in a complex set of small plumes. Similar buoyancy-driven instabilities can happen in disks, but there are many other instabilities that have been proposed as the source of the turbulence, including shear instabilities and instabilities springing from the joint effects of rotation and magnetic fields. Identifying instabilities is just the beginning of the story, though, as there is not yet any theoretical means for connecting instabilities to the amount of friction they cause.

Waves in the disk can also provide a means of transporting angular momentum without transporting mass. To take a terrestrial example, an undersea earthquake halfway across the globe can create a tsunami which threatens coastal California. The tsunami communicates energy and momentum to the places hit by the wave, but the mass of ocean water that hits the coast is not the same water the earthquake accelerates. The actual fluid motion in water waves is predominantly up and down, and does not transport much water horizontally. Rather, the pressure field set up by the earthquake accelerates the neighbouring water, which in turn stirs up a wave in the adjacent parcel of water, and so on until the wave encounters a barrier like a coastline where it crashes and dissipates, dumping its supply of momentum. Angular momentum transport by waves is a ubiquitous phenomenon in fluid dynamics. In disks, large scale spiral density waves can transfer angular momentum out of the inner disk, as can various waves involving magnetic fields.

Despite a lot of recent creative work on the subject, there is not yet a convincing resolution to the angular momentum problem, but observations definitely point to the operation of some kind of frictional process, even if it isn't yet clear precisely what that is.

Disk structure

The gravitational force on gas and dust of the disk is by far dominated by the gravity of the star, even in the outer regions of the disk. As indicated in Figure 7, the force of gravity on a bit of the disk, which always points in the direction from that point to the centre of the star, can be divided into a radial component parallel to the midplane of the disk and a vertical component perpendicular to the midplane of the disk. The radial component of gravity is mostly balanced by the centrifugal force of the gas swirling around the disk, although pressure gradients in the gas alter this balance in small but highly important ways. Both observations (including ALMA) and detailed computer simulations confirm that the centrifugal radius r_c sets the fundamental radial length scale for the disk. The centrifugal radius depends on the mass of the protostar (which determines the strength of its gravity) and the angular momentum per unit mass inherited from the portion of the Giant Molecular Cloud which collapsed to form the disk. The latter quantity can be expected to vary somewhat from one system to another, but a force balance shows that for any given value of the angular momentum per unit mass, r_c is inversely proportional to the mass of the protostar, simply because lower mass stars have weaker gravity and thus require less centrifugal force to balance it. Thus, all other things being equal, the disks around low mass protostars are spread out over a greater radius than disks around high mass stars. Recall, though, that the mass of a disk scales with the mass of the protostar, so that the density of gas and dust in the disk is generally lower for disks about low mass stars. Since it requires a certain minimum density for the disk to be visible in any given wavelengths, the two effects partly cancel, with the result that the apparent size of a disk is somewhat insensitive to the mass of the protostar. This effect can be seen in the ALMA observations in Figure 6, which shows disks about protostars ranging from .83 to 2 times the mass of the Sun.

The vertical component of gravity cannot be balanced by centrifugal force, which is purely radial. Left unbalanced, gravity would collapse the disk into a thin sheet, which to some extent it does—that is why protoplanetary disks are disk shaped. At some point, though, the pressure exerted by the compressed gas balances the force of gravity. The resulting characteristic thickness is known as the scale height, and is inversely proportional to the gravitational acceleration. Because both components of gravity become weaker as one moves away from the protostar, the scale height increases with radius in the disk, leading the disk to have a flared shape as shown in exaggerated form in Figure 7. For any given radius, the gas and dust density is highest at the midplane, and decreases with distance away from the midplane. The density of dust and gas at the midplane generally increases closer to the

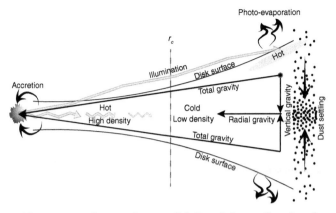

7. **The structure of a protoplanetary disk. Dust-laden gas flows into the outer edge of the disk from the surrounding Giant Molecular Cloud. As it moves inwards towards the protostar, conservation of angular momentum inherited from the Giant Molecular Cloud causes the swirl about the protostar to increase. The centrifugal radius, r_c is the radius at which the centrifugal forces from the surrounding swirl become comparable to the force of gravity pulling the gas and dust inwards. r_c is the fundamental length scale determining the extent of the disk. Frictional forces allow the gas and dust to continue to drift inwards and accrete onto the protostar. Dust settling is shown only at the outer edge of the disk, but in fact occurs throughout the disk. Other processes depicted in the sketch are discussed in the text.**

protostar, mainly because the gas and dust flowing through the disk is crammed into a smaller space as it approaches the disk centre.

The dust particles are much denser than the hydrogen gas, so it settles to the midplane just like a handful of dry dirt thrown into the air settles to the ground. The concentration of dust at the midplane is limited only by the turbulence which mixes it back away from the midplane, just as turbulence over a desert can maintain dust in suspension in a dust storm. The concentration of dust is very important, as high dust density favours the formation of rocky and icy cores.

The disk temperature decreases with radial distance because the flux of stellar energy decreases in inverse proportion to the distance squared, as energy is spread over ever larger spheres. Along the midplane of the disk, the decay of flux is even more pronounced, because the dust in the disk absorbs stellar radiation rather strongly, further increasing the temperature of the inner disk while decreasing that of the outer disk. Hydrogen at low pressure is fairly transparent to most of the energy output of a protostar, but it does strongly absorb ultraviolet and X-ray radiation. This is a small part of the total output, but the very energetic photons at these wavelengths are effective at blasting away hydrogen molecules back into the void, in a process called photo-evaporation. The flared shape of the disk is very important in this process, as it exposes the outer disk directly to UV and X-rays, as well as to the main energy output of the protostar, leading to hot spots near the disk surface (Figure 7). Photo-evaporation may be an important factor in mass loss from the disk. The phenomenon will make a reprise when we come to discuss loss of atmospheres from planets.

Recall that once formed, the mass of a protoplanetary disk does not grow with time. All mass accreted by the disk flows through the disk, and is accreted by the star, except for the small amount that goes into forming planets and other bodies of the eventual planetary system. The disk lives only as long as the net of gravitationally accreted mass and mass loss by photo-evaporation is positive. Once the balance goes negative, accretion onto the star clears out the inner disk, allowing the stellar illumination at the midplane, shown in Figure 7, to crash into the thick, opaque inner edge of the remaining disk, which rapidly evaporates the disk from the centre out. This is why disks live for a few million years, but then dissipate very rapidly.

Forming rocky cores

Dust grains in a protoplanetary disk start out as micron size. It is a long way from there to becoming a planet. Before you can make planets, you need to be able to make pebbles, and astrophysicists have been very clever at figuring out things that can go wrong on the way to that stage. Early on, gentle collisions allow grains to stick to each other and form millimetre- (mm) sized fluffy aggregates. So far so good, but at that point they hit the 'bouncing barrier' to further growth, because further collisions tend to fragment particles rather than aggregate them. Even if grains somehow surmount the bouncing barrier, they confront the drift catastrophe. Once pebbles reach centimetre (cm) size, then they orbit the star faster than the gas of the disk does. Gas swirls more slowly than a compact body in a vacuum would do because it is influenced by pressure forces as well as gravity. As pebbles become more massive, they are acted on more directly by gravity, and begin to orbit more like little planets. The pebbles thus feel a headwind which slows them down to the point where the centrifugal force is weaker than the force of gravity pulling them into the star. In the inner disk, all centimetre-sized pebbles would fall into the protostar within a century, if nothing opposed the process. There are so many known ways that planet formation can fail, it's unclear whether we would anticipate the existence of planets if we didn't know they exist.

The 'ugly fact' in this case is that in fact planets do form, so there must be a way around the pitfalls. The discovery of exoplanets, and the ability of ALMA to see the process of formation of planets in action in young disks, was a very big deal because it showed that planet formation is a robust process that happens around virtually every young star, and not just a result of some rare set of circumstances that prevailed in the Solar System. This is really more a matter of an ugly theory being slain by a beautiful fact, which forced scientists to keep thinking about the problem until they could find mechanisms that allow large planetesimals to form rapidly. This is an ongoing process, but the two most promising mechanisms are the streaming instability and pebble accretion.

Once the concentration of particles in a region becomes great enough for their self-gravitation to become important, the potent clumping effect of gravity, which we have already seen operating at many scales, can create planetesimals massive enough to resist the drag force that makes smaller particles fall into the protostar. The trick is to find a mechanism for concentrating the dust load sufficiently rapidly. The discovery of the streaming instability was an important theoretical advance providing a robust mechanism for planetesimal growth.

Gas–dust interactions are a peril to formation of planetesimals, but the streaming instability shows that the same processes can also be their saviour. Recall that dust generally orbits somewhat faster than gas in the disk. If there isn't much dust, then the force of dust pushing on the gas doesn't do much, but as the dust concentration increases, the dust begins to accelerate the gas. That means that the dust particles in that band experience less headwind, and drift inwards less rapidly. But then, dust drifting in from further out slows down as it encounters the accelerated-gas region, and accumulates there. But then the extra dust further increases the gas acceleration, leading to even more efficient trapping of dust, in a positive feedback loop. Eventually, the runaway growth in dust density allows self-gravitation to take over and build large planetesimals. The basic time scale for the instability to form planetesimals scales with the local orbital period (e.g. one year at Earth's orbit, twelve years at Jupiter's), and it takes about a hundred orbits for the instabity to concentrate dust to the point where gravitational self-aggregation activates. Thereafter, it takes only another ten orbits for the dust to clump into planetesimals of up to 100km in radius. All else being equal, the process would be expected to take longer as one goes further out in the disk, but the instability works better with bigger pebbles, so the availability of ice probably accelerates the process in the outer disk. In any event, one is able to form 100km planetesimals in a matter of some centuries. That's a good start.

The streaming instability still cannot account for the formation of Moon- or Mars-sized planetary embryos, let alone the rapid rocky core growth needed to trigger the formation of gas giants. For that, we turn to another important new idea—pebble accretion. This is another riff on the importance of gas–dust interactions. Let's step back a moment and consider how planetesimals might grow in the absence of gas. Imagine a bunch of 100km-sized planetesimals whizzing around. Most of the time they miss each other like ships passing in the night, but if two come close enough gravity will cause them to collide and often merge. Each planetesimal has an effective 'cross-section' which indicates the area it sweeps out within which collisions are likely. Because of the long-range effects of gravity, the cross-section is bigger than the physical cross-section of the solid body. The cross-section and planetesimal density gives the rate of encounters and rate of growth. Planetesimals can grow through collision and coalescence in the same way that cloud droplets grow into raindrops. In fact, for the most part, the equations governing the two processes are identical.

The problem, though, is that the collision cross-section is not really that much bigger than the geometric cross-section of the planetesimal, leading to a rather inefficient growth process. A mechanism is needed to greatly increase the cross-section for sweeping up mass. Enter pebble accretion.

Like the streaming instability, pebble accretion relies on the presence of gas. When a planetesimal is moving through a gaseous background it drags a bubble of gas with it. If a smaller particle (a 'pebble') whizzes by, it is caught in the gas drag of this bubble, and decelerated to the point where it is captured by the planetesimal's gravity. Simulations show that the effective cross-section for capture by pebble accretion can be hundreds of times the gravitational cross-section. Pebble accretion allows a planetesimal to cruise through the disk, hoovering up all available dust, with the process ending only when all the dust available in the feeding zone has been gobbled up. The process allows rocky cores of size sufficient to trigger gas giant formation to build up in a matter of some hundreds of thousands of years. Pebble accretion is not terribly sensitive to the density of the planetesimal, since most of the sweeping happens far away from the planetesimal surface where gravity only depends on total mass; it works as well for low-density icy bodies as for high-density rocky ones. In order to become efficient, though, the planetesimal must grow to a mass at least comparable to Earth's moon.

Pebble accretion can also occur in the inner regions of the disk, although probably at a slower pace. In the Solar System the process has led to the formation of inner rocky planets, one of which we call home. Gas giants need to form rapidly, before the gas in the disk dissipates, but rocky planets can continue to grow in size through collision of planetesimals for a hundred million years or more after disk dissipation. Simulations suggest this might have been the case for the inner rocky planets of the Solar System, but it should not be assumed that this is a typical outcome. The great diversity of architectures of planetary systems will be taken up in Chapter 5, and we will have the opportunity to revisit the issue there. Further, the place we see a planet today is not necessarily the place where it was originally formed. Planets can migrate through a number of mechanisms, offering up the possibility that planets can be manufactured out in regions of the disk propitious for their formation, and then migrate inwards.

Gravitational accretion of gas

Although the gravity in the disk is dominated by that of the protostar almost everywhere, sufficiently near a dense planetesimal the gravity of the planetesimal can dominate. So long as gas still exists in the disk, a planetesimal will attract gas within a certain radius, and so long as the body is dense enough that it is smaller than this radius, it will accrete gas, and dust with it. The gas cleared out from within the radius of attraction creates a relative vacuum, and new gas rushes in to replace it. There are two characteristic radii of accretion. The Hill radius is determined by the radius where the gravitational attraction of the planetesimal balances that from the protostar; within the Hill radius, the planetesimal gravity dominates. The Hill radius increases in direct proportion to the distance of the planetesimal from the protostar, and in proportion to the cube root of the ratio of the mass of the planetesimal to the mass of the protostar. For an Earth mass planetesimal in orbit at 5.2 au (Jupiter's orbital distance) about a Solar mass star the Hill radius is .05au. For the same planetesimal in orbit about a star with mass 20% of the mass of the Sun, the same Hill radius is achieved at an orbital distance of just 3 au, suggesting that effective gas accretion can occur closer to low mass stars than for high mass stars. For a given planetesimal mass, the Hill radius gets larger as one goes into the outer regions of the disk, favouring gas accretion, so long as one doesn't go so far out as to no longer have any gas available to accrete. The Bondi radius compares the rate of free-falling infall towards the planetesimal with the speed of sound in the disk, which imposes a kind of speed limit on how quickly gas can rush into the relative vacuum caused by accretion. The sound waves involved in this process are not different in kind from the sound waves we hear in air. Any gas can support such waves, with speed determined by properties of the gas. The Bondi radius is independent of the protostar's gravity, and so is not directly sensitive to either the stellar mass or the radial position of the planet in the disk. It is directly proportional to the mass of the growing planet, so, as expected, the accretion becomes more efficient for higher-mass planetesimals. The main dependence of Bondi radius on position in the disk comes from the temperature dependence of the speed of sound, which increases like the square root of temperature. Since the Bondi radius is inversely proportional to the square of the speed of sound, in the outer, colder portions of the disk, the Bondi radius is larger than in the inner hotter portions. In some circumstances, the Bondi radius controls gas accretion, while in others it is the Hill radius that dominates.

As gas accretes, it is compressed to relatively high density, and the increased mass leads to greater gravitational attraction of the surrounding gas, which in turn increases the rate of accretion. In its milder forms, this leads to exponential growth in the planet mass, but in its more extreme forms the rate of accretion grows so sharply with mass that the equations predict that that atmosphere mass would actually become infinite in a finite time—in fact, in times as short as eight million years. This, of course, cannot happen, but it is not entirely clear what arrests the accretion. It is not that the growing planet runs out of gas to accrete, because the flow of mass through the disk, which feeds the accretion of the protostar, is more than sufficient to feed the growth of the planet to the point where the planet itself would become a star. Evidently, the growing gas giant planet intercepts some of the inflowing gas, leaving the rest to accrete onto the protostar. This is a very active subject of research.

In the Solar System accretion of a massive gas envelope applies mainly to the gas giants Jupiter and Saturn, and perhaps to some extent to the ice giants Uranus and Neptune, which are less massive than the gas giants but still much more massive than the inner rocky planets. We shall see in Chapter 5 that amongst the exoplanets, it is common for much smaller planets to have accreted extensive gaseous envelopes.

Chapter 4
What are planets made of?

'We are such stuff as dreams are made on', wrote Shakespeare in *The Tempest*. The stuff of which we are made, our brains that can dream the luminous tales of Shakespeare, the food we eat, the air we breathe, the ground we stand on, all are forged in the interior of stars. This is true not just for us, but for other worlds in the Universe that have a chance to harbour life as we know it, and indeed probably also life as we don't know it.

Know your star

In order to understand how the elements are formed, it is necessary to know something about the life cycle of stars, and of the kinds of stars which populate the Universe. This information also provides essential background for most of the discussion of planetary characteristics in future chapters, since a planet's climate, the evolution of its atmosphere, and its ultimate fate are all linked to the properties and evolution of its host star.

Hydrogen being by far the dominant element in the Universe, the overwhelming majority of stars are made primarily of hydrogen, with a bit of helium thrown in, and are powered by the fusion of hydrogen into helium. These are called main sequence stars. For a main sequence star, it turns out that if you know its mass, you can deduce virtually everything else a reasonable person would want to know about the star and about the way it evolves over time. In essence, there is only one way to make a star with a given mass out of hydrogen, so all main sequence stars with the same mass are more or less alike. Main sequence stars range in mass from .08 Solar masses (the lowest mass that can ignite hydrogen fusion) to around 100 Solar masses. The corresponding properties are summarized in Table 1. Note that stars do not evolve along the main sequence. They are born with a certain mass, and remain on the corresponding part of the main sequence throughout their main sequence lifetime. The power output of a star is called its luminosity, and is typically measured in units of the Sun's present power output. Luminosity is a key stellar property, since it determines how much energy is available to heat any planets surrounding the star. Together with our distance from the star, luminosity also determines how bright a star appears in our night sky.

The surface temperature of a main sequence star, its radius, and its luminosity all increase with the mass of the star. Stars are categorized by their surface temperatures into the spectral classifications labelled O, B, A, F, G, K, M from hottest (and bluest) to coolest (and reddest). The Sun is a G star. Although more massive stars have more hydrogen to burn their greater luminosity overwhelms the fuel supply, and their main sequence lifetimes are sharply reduced relative to lower mass stars. The protostar in the disk HD163296 shown in Figure 6 is destined to become an A star, and whatever planets form around it will enjoy relatively clement main sequence conditions for under a billion years. More massive B and O stars go out in a blaze of glory after only some tens of millions of years or less. Our Sun will live on the main sequence for a total of ten billion years; we are a bit less than halfway through that time span now.

While a star is on the main sequence, it makes only helium. A star leaves the main sequence when it exhausts the supply of available hydrogen in its core. The post-main sequence fate of the star, and

Table 1. Characteristics of main sequence stars

Class	Surface temperature (K)	Mass	Luminosity	Lifetime	Fate
O	$\geq 30{,}000K$	$\geq 16M_\odot$	$\geq 30{,}000\mathcal{L}_\odot$	≤ 10 Myr	$^* \rightarrow SN$ $\rightarrow NS$ or BH
B	20,000–30,000K 10,000–20,000K	8–$16M_\odot$ 2.1–$8M_\odot$	$2{,}300$–$30{,}000\mathcal{L}_\odot$ 30–$2{,}300\mathcal{L}_\odot$	50–10Myr 800–50Myr	RG→SN→NS RG→WD
A	7,500–10,000K	1.4–$2.1M_\odot$	7–$30\mathcal{L}_\odot$	3–8Gyr	RG→WD
F	6,000–7,500K	1.04–$1.4M_\odot$	1.3–$7\mathcal{L}_\odot$	9–3Gyr	RG→WD
G	5,200–6,000K	$.8$–$1.04M_\odot$	$.4$–$1.3\mathcal{L}_\odot$	25–9Gyr	RG→WD
K	4,600–5,200K 3,700–4,600K	$.6$–$8M_\odot$ $.5$–$6M_\odot$	$.1$–$4\mathcal{L}_\odot$ $.05$–$1\mathcal{L}_\odot$	75–25Gyr 700–75Gyr	RG→WD sG→WD
M	3,400–3,750K 2,400–3,400K	$.2$–$5M_\odot$ $.08$–$2M_\odot$	$.006$–$05\mathcal{L}_\odot$ $.0005$–$.006\mathcal{L}_\odot$	2,000–700Gyr 6,000–2,000Gyr	sG→WD WD

Notes: The subscript $\odot$ refers to the Sun, so $M_\odot$ is the mass of the Sun, and $\mathcal{L}_\odot$ is its present luminosity. Luminosities are given at a stellar age midway through the main sequence. Myr stands for 'million years'; and Gyr stands for 'billion years'. The Lifetime column gives the approximate number of years the star spends on the main sequence before succumbing to its post-main sequence fate, given in the Fate column. Post-main sequence stage abbreviations are: sG = subgiant, RG = red giant, SN = supernova, WD = white dwarf, NS = neutron star, and BH = black hole. O stars leave the main sequence almost immediately; and depending on mass and composition pass through a series of transformations, indicated by **, too complex to summarize in this table. Stars that have a red giant stage pass through a subgiant stage beforehand. The final entry in the Fate column gives the kind of stellar remnant left behind after all fusion has ceased. Very low mass M stars do not go through a dramatic transformation as they age, but become gradually more luminous over trillions of years as they fuse hydrogen into helium, then fade into white dwarfs after exhausting their hydrogen fuel.

the range of elements produced, depends on the mass of the star. The lowest mass M stars, roughly below .2 Solar masses, are well mixed and can always bring hydrogen from outer layers to the core to sustain fusion. Such stars become gradually more luminous over trillions of years as they convert their hydrogen into helium, until they exhaust their fuel and fade away as white dwarf stars. White dwarf stars are no longer producing energy by fusion. Because of contraction, their surface temperatures are high (hence their bluish-white colour), but they have low luminosity because they are very small. White dwarfs are a common form of stellar remnant. All white dwarfs eventually cool down and go dark, but this process takes many times the current age of the Universe.

Higher mass M stars, and K stars up to .6 Solar masses, develop a shell of hydrogen fusing into helium surrounding the core when they leave the main sequence. These stars puff up into larger stars with relatively cool reddish surfaces at the end of their time on the main sequence. The luminosity of these stars—called subgiants—increases compared to their main sequence luminosity because the slightly cooler surface is more than made up for by the increase in surface area from which they radiate. However, their interiors do not become hot enough to produce much carbon, let alone heavier elements. Very low mass stars, such as K stars and M stars, thus do not contribute significantly to enriching the chemical composition of the Universe. Besides that, the low mass stars are so long-lived that the Universe is too young for them to have left the main sequence. The age of the Universe since the Big Bang is only 13.8 billion years, so the first M stars and most K stars still have not had time to leave the main sequence. Subgiants leave behind white dwarfs as stellar remnants after they exhaust their hydrogen fuel.

For more massive stars, up to eight times the Solar mass, the star first enters a subgiant stage similar to the evolution of lower mass stars, in which hydrogen is burned in a shell surrounding the core. Eventually, as the interior heats up, they enter a stage powered by a succession of fusion of heavier elements, in a process called nucleosynthesis. During this stage, the radius of the star swells immensely while the surface cools to the point where it emits a dull reddish light; hence, they are called red giants. First, helium is fused into carbon and oxygen, producing minor amounts of nitrogen as well. For sufficiently massive stars, carbon fusion then takes over, producing oxygen, neon, sodium, and magnesium, followed by neon fusion which produces more oxygen and magnesium. Then comes oxygen fusion, which produces elements between magnesium and sulphur, and finally silicon fusion which synthesizes (with varying degrees of efficiency) elements up through iron. Iron is the end state of fusion reactions, because elements heavier than iron require an input of energy in order to fuse, rather than releasing energy; elements heavier than iron release energy on fission rather than fusion, which is the principle on which fission power reactors work. Red giant stars blast off a great deal of the mass of their outer layers over the course of their life, efficiently dispersing the heavy elements they have synthesized into the galactic environment. I have glossed over some important differences between different kinds of red giant stars here, in order to provide just the essentials. When red giants that evolve from stars under 8 Solar masses exhaust their nuclear fuel, they contract and leave behind white dwarf stars as stellar remnants.

Stars with masses greater than 8 Solar masses go out with a bang, first passing through a red giant stage if their mass is below 16 Solar masses. For such stars, the core left behind after the mass loss of the red giant stage is so massive that rapid gravitational collapse ignites runaway fusion, leading to a stellar explosion known as a supernova. Stars more massive than 16 Solar masses also generally go supernova, although they first undergo a complex and varied sequence of transformations unlike the red giant stage of lower mass stars. More mass is lost during the supernova explosion, and the kind of stellar remnant left behind depends on the amount of remaining mass. If the remaining mass is 1.4 to about 2.2 Solar masses, it collapses into a neutron star. Neutron stars are so dense that electrons and protons have been squeezed together to merge into neutrons, which have no charge. A neutron star the mass of the Sun would be no more than 10 km in radius. The exact upper limit of mass for formation of a neutron star is uncertain, but above the estimated limit of 2.2 Solar masses, the remnant mass continues to collapse through the neutron star stage and forms a black hole. Black holes have such strong gravity that their interior is surrounded by an event horizon through which neither light nor matter can escape. Because of the complexity of the mass loss processes accompanying the various kinds of supernovae, it is difficult to determine the range of main sequence masses that leave behind neutron stars or black holes as remnants. Any star that goes supernova could potentially leave behind a neutron star. As a rough guide, though, stars with 8–25 Solar masses when they leave the main sequence most probably leave behind neutron stars, whereas more massive stars are more likely to leave behind black holes. The mass thresholds also depend somewhat on the initial composition of the progenitor star.

Where do elements heavier than iron come from? Red giant stars can build some heavier elements through the 's-process' (slow process) form of nucleosynthesis which involves adding mass to the elements already synthesized by capture of neutrons. However, in a red giant, the supply of neutrons is feeble, and the process is inefficient. Stars more massive than about 8 Solar masses ultimately end their lives as core-collapse supernovae, although some of them first pass through a red giant stage. When fusion becomes insufficient to support the core against gravity, rapid collapse leads to a stellar explosion. After things settle down, core-collapse supernovae leave behind neutron stars or black holes. During the supernova explosion, a storm of neutrons is produced, many of which are captured by the elements up to iron present in the star. These add mass to the lighter elements, allowing all the heavier elements to be produced, either directly or as fission products of unstable nucleii. This is referred to as the 'r-process', or rapid neutron capture-process. Supernovae in the early Universe can even start with just hydrogen and helium and synthesize nearly the full range of heavy elements found in nature, up through uranium. Some elements with lower atomic mass than iron, which can be produced by the s-process, have their primary source in the more efficient r-process synthesis in supernovae. This is the case for potassium, which is crucial to life as we know it.

Very recent observations of neutron stars made possible through the breakthroughs in gravitational wave detection—waves in the fabric of spacetime itself—have revealed a startling new source of r-process elements. The neutron stars left behind as stellar remnants of sufficiently massive stars sometimes appear in pairs orbiting each other. Binary neutron stars eventually undergo cataclysmic merger, and the resulting maelstrom of pure neutrons provides an ideal environment for synthesis of elements by neutron capture; the cloud of r-process elements left behind by merger has been observed, and suggests that neutron star merger may be a more important source of r-process elements beyond iron than core-collapse supernovae. But new advances in the understanding of the sources of heavy elements are proceeding at a breathtaking pace. There are now indications that yet more exotic events, called collapsars, may be the dominant source of the heaviest elements. Collapsars are whirling maelstroms that form around a black hole when very rapidly rotating extremely massive stars collapse. They provide a neutron-rich environment ideal for breeding the heaviest elements.

The ingredients most essential for life—hydrogen, oxygen, nitrogen, and carbon—plus the rock-forming materials needed to make a decent rocky planet for it to live on are all present in the elements up through iron, so there would appear to be no pressing need for a habitable Universe to contain any element heavier than iron. Tiny amounts of certain neutron-capture elements—nickel, copper, zinc, and molybdenum—are crucial to the functioning of enzymes that power the metabolisms of life as we know it, but at present it is unknown whether evolution could produce functionally equivalent enzymes in the absence of these elements. A universe without neutron-capture elements might in some ways be a safer universe, as it would be one without nuclear weapons (even hydrogen bombs need a uranium or plutonium fission igniter). However, we'll see in Chapters 6 and 7 that long-lived radioactive elements, including uranium, thorium, and a radioactive form of potassium, are crucial to maintaining the habitability of planets over billion-year time scales.

Although the first stars of the Universe could not have formed planetary systems, the process did not take long to get underway. Because massive stars are short-lived, the first billion years of the Universe already had time for 1,000 generations of production of the heavier elements. Observations show that the early universe was already a quite dusty place. Although massive stars do not live long enough to host planetary systems where life is likely to emerge, they are essential to producing the elements that lower-mass systems use to build habitable worlds. The Milky Way galaxy, our home, formed not long after the Big Bang, and has been building its stock of heavy elements ever since. Most of this galactic chemical evolution remains internal to the galaxy, although galaxies do sometimes collide and exchange material. Over the past thirteen billion years of nucleosynthesis in the Milky Way, there has been ample time for thorough mixing across the galaxy. Thus, our Solar System incorporates ingredients from a mix of myriad expired stars, most of which have been processed multiple times through short-lived stars. Every breath you take includes oxygen atoms from thousands of different stars that have lived and died in our galaxy over the past thirteen billion years.

Before leaving this digression on the nature of stars, we need to take a look at how their luminosity evolves over time. For the most part, planet formation plays out before the host star enters the main sequence, and it is important to know the luminosity of the star during its early life because that sets the temperature of the disk, which in turn affects many aspects of planet formation.

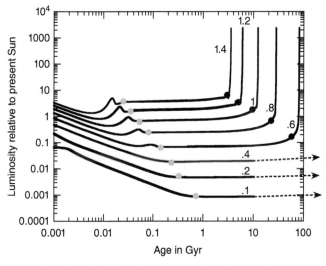

8. Luminosity vs. time for main sequence stars of masses from .1 to 1.4 Solar masses. Luminosity is measured relative to the present luminosity of the Sun, so a star with luminosity 1 on this graph has the same power output as the present Sun. Time is measured in billions of years. Note that both axes are logarithmic, in order to allow the enormous range of age and luminosity to be clearly displayed. Grey circles mark the approximate time of entry onto the main sequence, following ignition of hydrogen fusion. Black circles indicate the end of the main sequence, which for the more massive of these stars is followed by a helium flash and entry into the red giant stage. The stars of mass $.4M_\odot$ and below do not leave the main sequence until hundreds of billions, or even trillions, of years past the right-hand edge of the graph.

Figure 8 shows the luminosity evolution over time, from the earliest stages of stellar collapse, on to the main sequence and beyond. In this chapter, we will only be concerned with the first ten million years of the evolution at the left of the graph; the evolution during the main sequence and dramatic events at the end will be taken up in Chapters 5, 6, and 7. The main result of relevance to the rest of this chapter is that for stars of $.8M_\odot$ or less, the luminosity at the beginning of the T-Tauri stage is considerably greater than it is once the star settles down onto the main sequence. The effect becomes stronger as the mass of the star is reduced. For example, for a red dwarf star of a tenth the mass of the Sun, the luminosity at the beginning of the T-Tauri stage is a hundred times greater than it is once it enters the main sequence. Thus, the planets we see today illuminated by faint low mass stars were born in a disk heated by a much brighter protostar. Since low mass stars, being long-lived and not taking much mass to form, are by far the most common kind of star in the Universe, these are quite important planets to think about.

Solar composition

Since almost all of the mass that collapsed to form the protosolar disk made its way into the Sun, the composition of the Sun provides a good indication of the raw materials from which the Solar System was built. Recall that the Sun is not at present synthesizing anything other than helium, so apart from helium the composition of the Sun hasn't changed much over time. By an arduous analysis of spectral lines, the elemental composition of the surface of the Sun has been quite well characterized, and while models indicate that the surface composition should differ somewhat from the mean composition of the entire Sun, the differences are small enough that the surface composition is quite representative of the bulk composition. Not all the stable elements can be detected, since some lack suitable spectral lines, but almost all can. The result is shown in Figure 9, where abundance is plotted against the atomic number of each element. The atomic number is the number of protons in the nucleus, and largely determines the chemical properties of the element. There is a general abundance trend downwards with increasing atomic number, but relative to this trend iron (Fe) is anomalously abundant because it is the endpoint of fusion nucleosynthesis.

With the exception of hydrogen, nucleii need to incorporate a number of neutrons roughly equal to the number of protons in order to be stable. The total mass of neutrons and protons is called the atomic mass, and is shown in the inset of Figure 9. Most elements can exist in multiple stable or long-lived forms with different numbers of neutrons. These variants of an element, which are chemically similar, are called isotopes. Nucleosynthesis almost invariably prefers one dominant isotope. For example,

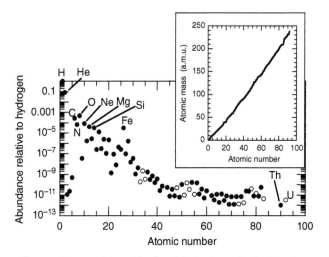

9. **Elemental composition of the Sun. The horizontal axis gives the atomic number (number of protons) for each element and the vertical axis gives the ratio of number of atoms of the element to atoms of hydrogen. Some elements cannot be measured in the Solar surface with present technology, and these have been filled in from estimates based on composition of primitive meteorites (open circles). The remaining missing elements, e.g. elements 84–89, are radioactive elements with short half-lives, which are expected to have decayed away. The inset shows the atomic mass of the elements, in atomic mass units (a.m.u., approximately the mass of a proton). The atomic mass is essentially the number of protons + neutrons in the nucleus of an atom.**

carbon occurs in a stable form with atomic mass 12 and atomic mass 13, and the first is by far the most common in Solar composition. The atomic masses shown in Figure 9 are very nearly those of the most common isotope of each element.

Astronomers refer to any element heavier than helium as a metal, even though this doesn't correspond to the chemists' definition of a metal. The metallicity of an object (typically a star) is defined as the ratio of the mass of everything other than hydrogen and helium to the total mass. The photospheric metallicity of the Sun is 1.34%, and the estimated corresponding bulk metallicity is 1.4%. This is broadly consistent with the percent-level concentration of dust in observed protoplanetary disks.

Nearby star-forming regions to the Solar System show similarly high metallicities as the Solar composition, and loosely speaking this appears to be the case throughout the flat spiral arm region of the Milky Way and of nearby spiral-arm galaxies. There is, however, a population of very old stars elsewhere in the galaxy that has considerably lower metallicity than Solar composition, having metallicities as low as 0.3%. These are stars that formed early in the history of the galaxy, before many heavy elements had been synthesized. Elliptical galaxies, which do not have spiral arms and have lower star-formation rates, are largely composed of such stars. However to date no primordial stars with essentially zero metallicity have been found.

Volatility

A chemical substance can exist in various phases, the most familiar of which are solid, liquid, and gas. They are distinguished from each other by density, the arrangement of molecules, and various other thermodynamic properties. Liquids can play a role in the chemical evolution of the disk, but here we'll limit attention to the solid-gas phase transition. For the solid form to exist, the temperature needs to be below the freezing point for the substance in question; for the low pressures present in the disk, the freezing temperature can be considered independent of pressure. For example, the freezing point of metallic iron is 1,811K (1,538°C), of water is 273K (0°C), of carbon dioxide is 194K (−78°C), and of carbon monoxide is a chilly 68K (−205°C).

Even below the freezing point, molecules of the solid are continually flying away from its surface and entering the gas phase. The rate at which this happens increases very rapidly with temperature, and if the escape isn't compensated by molecules of the same substance colliding with the solid surface, the solid will shrink and disappear into the gas phase in a process called sublimation. For liquid converting to gas, the analogous process is called evaporation. The rate at which gas phase molecules hit the solid surface is proportional to the pressure exerted by the molecules of that same substance that are present in the surrounding gas, and, at a given temperature, every substance has a characteristic pressure that will maintain it in equilibrium called the saturation vapour pressure. The moisture in your breath condenses on a cold winter day because the saturation vapour pressure in the cold air with which your breath mixes is much lower than the saturation vapour pressure in the warm, moist interior of your lungs. The saturation vapour pressure is always a very strongly increasing function of temperature, but its value at a given temperature varies greatly amongst substances. Substances that require a low temperature to avoid sublimating away are called 'volatile', from the Latin root which means 'to fly away'. In French, the word is also used to refer to domestic birds, and the molecules of a volatile solid do indeed fly away into the void when they reach moderate temperatures. Substances which can survive high temperatures in the disk without sublimating are called 'refractory'. The dividing point between the two classes is arbitrary, and there is in fact a continuum of volatility between the two extremes. Because temperature in the protoplanetary disk decreases as one moves outwards from the protostar, volatility determines which substances can be present in the condensed phase in various portions of the disk. These are the substances which make up what we have been loosely referring to as 'dust'. In practice, the freezing point of a substance gives a pretty good indication of its volatility.

Some elements in the gaseous disk can condense into their solid elemental form, such as iron condensing into metallic iron or carbon condensing into graphite grains, but in the relatively high-pressure environment of the disk atoms collide frequently enough for chemical reactions to proceed rapidly, forming a great variety of molecules. The volatility of a molecule differs greatly from that of the pure elements that make it up. For example molecular hydrogen is so volatile that it can't condense anywhere in the disk, and molecular oxygen O_2 has a freezing point of 54K which means it is also extremely volatile. However when the two react to form water (H_2O), the freezing point increases to 273K and the result is much less volatile.

Complex thermodynamic calculations are needed to determine the precise mix of molecules that condense out of the gaseous disk, but we can get a good indication of what can be made by looking at the relative availability of elements in the Solar composition, which is in turn probably a pretty good indication of what's available to make planets with in nearby planet-forming regions, and perhaps farther afield in the spiral arms of galaxies similar to our own. After hydrogen, the most abundant elements by count of atoms are helium (He), oxygen (O), carbon (C), neon (Ne), nitrogen (N), magnesium (Mg), silicon (Si), and iron (Fe). Helium and neon are examples of noble gases, which essentially don't react chemically and are also extremely volatile. Among the rest, the relatively abundant oxygen reacts with the essentially unlimited supply of hydrogen to make water (H_2O), carbon reacts to form methane (CH_4), and nitrogen reacts to form ammonia (NH_3), all of which are moderately volatile, with water being the least volatile of the three. Magnesium, silicon, and iron do not easily bond to hydrogen at low pressures, so these elements typically react with the less abundant higher atomic number species, although iron can also appear unreacted as metallic iron. These compounds are invariably very refractory and can survive the high temperatures of the inner disk. The silicates are a particularly important family of minerals. These consist of the silicate ion (SiO_3) or its close relatives bonded to a heavier element. Enstatite ($MgSiO_3$) can be made out of some of the most abundant elements in Solar composition. Silicates have low volatility. For example, enstatite freezes at 1,830K. Silicates are a vast family of minerals, and can contain iron, calcium, aluminium, and various other metallic elements in various combinations. One member of the family of silicate minerals called olivines is $MgFeSiO_4$. Silicates, together with metallic iron, make up the basic stuff of rocky planets and the rocky cores that accrete gas to form planets with an extensive gaseous envelope, including gas and ice giants.

Most of the mass of an atom is in its nucleus, which takes up very little space. Although atoms don't have a sharp boundary, their size is determined by the extent of the much larger electron cloud surrounding the nucleus. For example, the diameter of a hydrogen atom is over 60,000 times the diameter of the proton that makes up its nucleus. The atomic number determines the number of electrons surrounding the nucleus, which affects the size of the atom. The atoms of high atomic number elements are only somewhat bigger than low atomic number ones, and so they can pack more mass into a smaller volume, especially in view of the fact that atomic mass tends to increase with atomic number. The connection between atomic number and the size of the atom is not strict. Although the higher nuclear charge at higher atomic number does tend to pull electrons inwards, the electron configuration can have a significant effect on atomic radius, which can offset the mass and charge increase of the nucleus. Nonetheless, solids composed of molecules of high atomic mass elements tend to be denser than those made of low atomic mass elements. Enstatite, made of silicon and oxygen, is about three times as dense as water ice which is made of low atomic mass hydrogen and oxygen. Carbon dioxide ice is 1.6 times the density of water ice, because of the greater atomic mass of carbon and oxygen as compared to hydrogen, Solid iron is about twice as dense as enstatite. What we commonly refer to as 'rocks' are really just high molecular mass ices. They also tend to be much less volatile than low molecular mass ices. High-density planetesimals are also more tightly bound gravitationally, because the force of gravity decreases with distance and the parts of a high-density object are closer to each other than is the case for a low-density one. High-density planetesimals are thus more resistant to fragmentation.

Where has all the carbon gone?

Rocky planets can have very different composition from the protoplanetary disk, and are expected to be particularly depleted in more volatile elements. There is at present scant information about rocky planet composition for exoplanets, but very clear evidence from the Solar System. This evidence comes from the composition of Earth's silicates and from primitive meteorites known as CI chondrites, which provide some indication of what protoplanets were made of. To judge from the Solar composition, Earth should be a carbon-dominated planet rather than a silicon-dominated one, since carbon is much more abundant in Solar composition than silicon. Earth's rocky component is 20% silicon but under a tenth of a percent carbon. Some of that missing carbon could have dissolved in the metallic core, but it is hard to account for much of the disparity that way. CI chondrites are also very depleted in carbon relative to the Solar composition, although not so much so as Earth. The mystery is compounded by the fact that elemental carbon is not very volatile. Evidently, at some stage in the process of forming the Earth and other rocky inner Solar system bodies, carbon was bound up into volatile CO, CO_2, and CH_4, which are especially easy to lose to space when planetesimals are small and have weak gravity. It is fair to say that nobody really has a clear idea of where the carbon went. Information about rocky exoplanet composition would be a great help in resolving this problem, but while the quest for such information is an active research topic, it will be hard to come by.

The snowline

One of the critical characteristics of a disk is the snowline—the distance from the protostar where the disk gets cold enough for ice to form. Outside the snowline, the formation of ice can greatly accelerate the formation of planetesimals. Each form of ice has its own snowline, but since water is generally the most abundant ice-forming molecule, and one with the highest freezing point among low molecular mass ices, there is a particular interest in the snowline for water ice.

The temperature of a chunk of ice in the disk is determined by a balance between the energy absorbed from the protostar's light and the energy radiated as infrared. Clean snow or ice is highly reflective, but the ice that forms in the disk is quite dirty because of its dust content. As any dweller of a cold Northern city knows, it takes rather little contamination to turn snow nearly black. Recalling the blackbody radiation formula, the energy radiated as infrared increases like the fourth power of the temperature of the body. If the protostar light were propagating in a vacuum, its flux would decrease inversely with the square of distance. When the temperature much exceeds the freezing point, ice will quickly melt and evaporate back into the disk. The energy balance implies that the snowline position increases like the square root of the stellar luminosity, and decreases in inverse proportion to the square of the freezing temperature for the substance in question. For a Sunlike protostar, the freezing point for water ice would be around 3.6au, putting the present orbit of Jupiter well outside the snowline. For a low mass M star like Trappist 1, which has a mass a bit under a tenth of a Solar mass, the snowline for the protostar would be considerably closer, at a mere .5au for water ice. For CH_4 ice, the snowline is at 32au for a Sunlike protostar and 4.6au for the low mass case.

In reality, the snowline positions would be closer to the protostar than the estimates given above, because of attenuation of the protostar radiation by dust absorption near the midplane of the disk. Nonetheless the vacuum estimates give a reasonable indication of where the snowline is.

In the prevailing view ice giants like Uranus and Neptune are dominated by low molecular mass ices. If this view is correct, they could not be built through gas accretion since the gas is by far dominated by hydrogen. Gas accretion could play a role in the proportion of hydrogen in ice giants, but the most plausible explanation for ice giants would be that they were built by accretion of icy planetesimals. However, we'll see in Chapter 5 that amongst exoplanets there are many warm Neptune-sized low-density planets that appear to have a much higher hydrogen content than Uranus and Neptune. Even for our own Uranus and Neptune, the conventional view of their dominant icy composition has been called into question, as many of their characteristics could also be accounted for by a more substantial rocky core surrounded by a primarily H_2/He envelope. The mix of rock, ice, and H_2/He that goes into the recipe for Uranus and Neptune is currently an active area of research, and will only be resolved by future missions to those planets aimed at providing more information about their interior structures.

Core segregation: a special role for iron

In the era of exoplanets, astronomers are learning the importance of distinguishing, among the things they have traditionally called 'metals', between actual metals and rocks, which are mostly silicates. The importance of the distinction has long been recognized in the Earth science disciplines. Among true metals, iron plays a special role, because of its anomalously high abundance.

When a planetesimal becomes big enough to have significant self-gravity, heavier constituents can sink to the centre, leading to chemical differentiation of the body. This can happen both in the planetesimal stage and when planetesimals aggregate by collision to form planets, and it is generally thought that efficient differentiation requires the body to be in a molten state—i.e. that it has a deep magma ocean. Magma oceans can be formed from the heat of collision, or from the heat released by decay of short-lived radioactive elements. Some iron goes into silicate minerals, but much of it forms dense metallic iron droplets that sink to the centre of the body and form a metallic core. This has happened in all of the rocky bodies of the inner Solar System, including the Moon.

When a metallic iron core segregates near the centre of a planet, it leaves behind a surrounding layer of silicate minerals known as the silicate mantle. The minerals that make up the mantle are partly, although not completely, depleted in iron. Formation of a metallic iron core significantly increases the average density of a planet. Iron has high atomic mass, but is also a small atom so it takes up a lot less space when it is stripped out of a silicate molecule and packed into a volume of pure iron. For example, consider the family of minerals called olivines, which make up most of the Earth's upper silicate mantle. The family has two end members, Mg_2SiO_4, called forsterite, which has a density of $3,300kg/m^3$, and Fe_2SiO_4, called fayalite, which has a density of $4,400kg/m^3$. Metallic iron, with a density of $7,800kg/m^3$, is much denser than either. If a mix of metallic Mg with fayalite is chemically transformed into a mix of metallic Fe and forsterite, the density of the mixture increases from $3,400kg/m^3$ to $4,400kg/m^3$. All these densities are given for the solid form at room temperature and low pressure, so in an actual planet they need to be adjusted for density differences in the liquid phase, and due to compression in the planetary interior, but the general trend remains valid.

The chemical effects of core segregation are even more important than its effect on density. In particular, iron loves to react with oxygen, depriving other molecules of their chance to pair up with oxygen. Thus, if there is a lot of iron around in the silicate mantle, non-oxidized species such as H_2 and CH_4 tend to be outgassed into the atmosphere by volcanism, whereas if metallic iron has segregated into a core the outgassing has a higher proportion of oxidized species, such as CO_2. In addition, significant amounts of important planetary constituents such as carbon, nitrogen, and sulphur can dissolve in the liquid core and be permanently sequestered there. The heavy radioactive elements like uranium and thorium have very low solubility in the core, and so mostly remain in the silicate mantle. Potassium, including its long-lived radioactive isotope potassium-40, also stays mostly in the silicate mantle.

Most thinking about core segregation on rocky planets has been directed at the Solar System, but differentiation could occur very differently on rocky exoplanets that are bigger, or hotter, or have different chemical composition from what we know in the Solar System. It is possible that in some cases the familiar Earthlike differentiation doesn't occur at all. Planetary scientists are still struggling to grasp the vast array of new possibilities offered by exoplanets.

The volatile envelope

For gas giants, there is no mystery as to where their atmospheres came from. They are practically all atmosphere, sucked in from the hydrogen-dominated gas and ice of the ambient disk. Ice giants are also not problematic, as they are built largely from icy planetesimals, with some admixture of ambient gas. Accounting for the presence and composition of the atmospheres of rocky planets, and also of the numerous planets intermediate between Earth-sized rocky planets and Neptune-sized low-density planets found amongst the exoplanets we shall meet in Chapter 5, presents a considerably greater challenge.

A primordial atmosphere is the atmosphere a planet winds up with as it approaches its maximum mass at the end of the planet formation stage. There are two ways a planet can get a primordial atmosphere. If the rocky or icy core of the protoplanet becomes sufficiently large while there is still abundant gas in the disk, its gravity can attract enough gas to form an atmosphere. Such an atmosphere would be primarily H_2. If the rocky cores grow slowly in mass, they can accrete a lot of gas without passing the threshold that triggers runaway gas accretion and turns them into gas giants.

The second way to get a primordial atmosphere is to cook it out of rocky or icy planetesimals from the heat of collision as they aggregate to form the planet. This would yield an atmosphere rich in the higher molecular weight volatiles, such as water, but the energy of the collision could still strip hydrogen out of the hydrogen-bearing volatiles. This can produce an H_2-dominated primordial atmosphere even though planetesimals contain very little pure H_2.

When planetesimals that are small enough to have weak gravity collide, the resulting heat evaporates volatiles which can then escape to space. For the Solar System we can directly see the evidence of this process by comparing ancient primitive meteorites with the Solar composition. Planetesimals in the hot inner disk may also be born volatile-poor, because of the inability to accrete volatiles in the form of ice. Nonetheless, judging by primitive meteorites, the raw material of rocky planets still can offer a good supply of volatiles. CI chondrites contain very little ice but do contain hydrated minerals that store significant amounts of water in their crystal structure. They have been found to contain 10–20% water by weight, which is far more than sufficient to supply the Earth's oceans. In fact Earth's oceans make up only .02% of the mass of our planet, and even allowing for possibly another two or three oceans' worth of water stored in hydrated minerals in the Earth's interior, Earth begins to look like a rather dry planet. If we had retained even half the water in planetesimals with composition similar to CI chondrites, our continents would all be deeply submerged.

Even after a planet reaches its mature size, its atmosphere is continually leaking away to space to one extent or another. Atmospheric escape plays an important role in sculpting a planet's atmosphere and even the chemical composition of its rocky component. For a bit of atmosphere to escape the planet's gravity well, it must be boosted to above the planet's escape velocity, much like a rocket, and that takes energy. In this case, the rocket fuel is generally provided by the absorption of energetic ultraviolet light from the star high up in the atmosphere, and by the energy of stellar wind particles interacting with the upper atmosphere. The former process is called photo-evaporation, and the latter stellar wind erosion. For fixed density, the escape velocity increases in direct proportion to a planet's radius, and in proportion to the square root of its mean density. For Earth, with a mean density of $5,500 kg/m^3$, the escape velocity is 11km/s. Jupiter, being made mostly of H_2 has a mean density of $1,330 kg/m^3$—just a bit more than water—but because its radius is eleven times that of Earth its escape velocity is 59.4km/s. It takes a very large input of energy to sustain appreciable atmospheric escape from a gas giant, but for gas giant exoplanets orbiting very close to their host stars where they can receive a lot of energy, it can happen.

Only a small part of the energy a planet receives from its star is available to sustain escape. Except for extremely hot planets, the light from the star absorbed deep in the atmosphere goes almost entirely into creating infrared radiation which escapes to space, rather than accelerating molecules which then carry their energy (and mass) away from the planet. Shortwave ultraviolet and X-rays, as well as stellar wind energy, are deposited very efficiently in the thin upper parts of the atmosphere where molecules have a chance to escape before colliding with another molecule, or where the intense heating of the low-density gas there can drive an organized outward flow of mass. The latter form of escape is called hydrodynamic escape, and is only triggered when the energy received exceeds a certain threshold which depends on the escape velocity and the mean molecular mass of the upper atmosphere. When the threshold is exceeded, the mass loss can be extreme, involving an outflow that achieves supersonic speeds. Planets orbiting M stars are especially vulnerable to atmospheric escape, because M star luminosity is so high during the long pre-main sequence stage and also because M stars put out a high proportion of energetic UV radiation throughout their life, as compared to Sunlike G stars.

Low molecular mass constituents escape more readily than high molecular mass ones. When molecules are escaping individually, this is because it takes less energy to accelerate a low mass molecule to escape velocity than a high mass one, for much the same reason that it is easier to toss a golfball onto a roof than to get a bowling ball up there. Hydrodynamic escape takes the form of an organized continuous stream of outflow rather than individual molecules escaping, but molecular mass still works its way into the escape criterion, because of its effect on heat capacity. A kilogram of H_2 has the same gravitational potential energy binding it to the planet as a kilogram of N_2, but at any given temperature contains about thirteen times as much thermal energy that can sustain escape. Thus, loss of H or H_2 is especially favoured, and if the mass loss is vigorous enough, the escaping hydrogen can drag some heavier molecules along with it.

Even if a planet is stripped to its rocky core through erosion of its primordial atmosphere, there are various ways it can regenerate a secondary atmosphere. Long after the disk has cleared of gas, it can obtain a new supply of volatiles through bombardment by remaining swarms of volatile-containing rocky planetesimals, or through bombardment by icy bodies such as comets that sometimes get flung inwards from beyond the ice-line. This scenario is known as the Late Veneer. In addition, in the course of formation the silicate interior of a rocky planet can sequester a considerable quantity of volatiles. Silicate mantles are solids in the sense that their molecules are locked in a crystal structure, but they are hot enough that they can still flow like a very viscous fluid, just as glaciers can flow even though they are made of solid ice. Except for very young planets, the heat that drives the flow of mantles generally comes from radioactive decay of unstable elements by fission. The short-lived radioactive elements decay quickly, but others such as uranium-238, thorium-232, and potassium-40 can provide heat for billions of years. Mantle flow brings hot material towards the thin crust, where the low pressures allow it to melt. For most materials, squeezing hard favours formation of solids, water being a notable exception. The melt reaches the surface in the form of volcanoes, which release gases into the atmosphere. The composition of the outgassing depends on the mantle chemistry, temperature, and pressure. For conditions similar to the Earth's present mantle, the outgassing mostly contains oxidized gases such as H_2O, sulphur dioxide (SO_2), and CO_2. Before the core had segregated, non-oxidized ('reduced') gases such as H_2, CH_4, and hydrogen sulfide (H_2S) would be favoured. This stage is believed to have been very short on Earth, but could be longer on other planets. Reduced secondary atmospheres could also be formed by late-veneer impacts of planetesimals without a differentiated iron core, or planetesimals which melt and mix upon impact. Exoplanetologists are struggling to grapple with the implications of a wide range of possible alien geologies. For example, on a planet with a higher ratio of C to O, it might not be possible to form CO_2 even after iron had separated into a metallic core, and carbon might come out in the form of CH_4 instead. The carbon might also remain in the form of graphite, or get bound up in silicon carbide (SiC), so that not even CH_4 would outgas. Or, on a larger rocky planet than Earth, the higher subsurface pressure makes it harder to form melts, which suppresses volcanism and outgassing, and even if melts do form, they may be negatively buoyant and sink, rather than rising in volcanic plumes as happens on Earth.

Some gases, such as neon, which are abundant in a Solar composition disk, are noble gases which do not form chemical compounds under natural circumstances. They are also very volatile, so can only be delivered in gaseous form. This is probably why Earth has very little neon in its atmosphere. The relative absence of neon indicates either that Earth never accreted a primordial hydrogen atmosphere, or if it ever had one lost it by such vigorous means as to be able to drag almost all the heavy neon with it. Molecular nitrogen (N_2) is not a noble gas, but it is almost noble in that the molecular bond is very strong and hard to break. N_2 does not easily form minerals the way carbon does. Earth and Venus are very depleted in nitrogen relative to Solar composition, but N_2 makes up most of our atmosphere, and certain microorganisms have evolved to be able to break the bond and make nitrogen available for life. The factors governing how much nitrogen a rocky planet gets are still rather mysterious. It is an important question to resolve, as nitrogen is an essential ingredient for life as we know it.

Regardless of what is outgassed from the interior, the resulting atmosphere is a dynamic entity. The source of atmospheric constituents from outgassing is balanced by the sink due to escape to space and chemical reactions with the planet's crust. This balance determines the long-term chemical evolution of the atmosphere and mantle.

We say that a planet has a distinct surface, such as the rocky surface of Earth's continents or the whole of the Venus or Mars surface, if there is a sharp jump in density when crossing some boundary. As noted before, rock is just a form of high-temperature ice, so there is no fundamental difference between a solid silicate rock surface or a solid water ice surface, save for the range of temperatures and pressures at which they can exist. The surface could also consist of a discontinuity between a dense liquid phase, such as Earth's ocean, and a less dense gas. For a planet where some substance (e.g. water) condenses into a dense liquid phase somewhere, an ocean would form if the cloud droplets—which fall because they are denser than their surroundings—remain liquid indefinitely as they descend, so that they accumulate at depth and aggregate into a liquid layer with a sharp density jump at the surface. An ocean can be thought of as just a high-density extension of the atmosphere. A typical Earthlike planetary layered structure with a thin atmosphere and ocean is shown in Figure 10(a). 'Thin' in this context means that the mass of the atmosphere and ocean make up a very small proportion of the total mass of the planet, and that they account for just a small part of the radius. Venus would count as having a thin atmosphere even though the mass of its atmosphere is nearly a hundred times that of Earth. Rocky planets with oceans can also have a continental crust that rises above the ocean surface, and the oceans and continents could be partially or totally covered by a thin ice layer. Rocky planets can also manifest as bare rockballs without atmosphere or ocean.

At high pressure, liquids can freeze into a great variety of solids that cannot exist at low pressure. In the case of water, for example, liquid freezes into conventional low-pressure ice—Ice I—at temperatures below 273K. Ice I floats on liquid water. However, at pressures ten times those found at the bottom of the Earth's ocean, Ice VI forms, which is denser than the liquid and stays at the bottom. Ice VI and other solid phases can exist at temperatures of 600K or more. In Kurt Vonnegut's novel, *Sirens of Titan*, Ice IX is claimed to catalyse freezing of the ocean at room temperature. There really is an Ice IX, but it has no such properties, and requires high pressures to exist. At very high pressures, ice can even take on the form of an exotic semi-liquid, electrically conducting phase called superionic ice. For a planet composed largely of water, such high-pressure ice phases would form an ice mantle which would take the place of Earth's silicate mantle (Figure 10(b)).

Sharp phase boundaries need not exist anywhere in a planet's structure, and the conditions under which they form depend on the temperature and pressure profile of the planet with depth, as well as its composition. In Jupiter, or indeed any gas giant whether hot or not, hydrogen just gets gradually denser and more liquid-like at greater depths, and there is no distinct hydrogen–ocean surface (Figure 10(d)). If Earth's surface temperature were made greater than 647K (374°C), water would behave similarly, with the former ocean shading continuously into the atmosphere, with no sharp boundary in between (Figure 10(c)). Although there would be no meaningful distinction between ocean and atmosphere in a case like this, at higher, colder altitudes there would still be liquid water clouds forming, which would have a density discontinuity between the liquid drops and surrounding gas. As these drops fell into hotter regions, they wouldn't accumulate into an ocean, but instead their density contrast would just fade to zero and they would merge smoothly into the high-density fluid below. Every substance has its own *critical temperature* above which the liquid–gas transition disappears. For carbon dioxide, the critical temperature is just 304K (31°C), delineating the maximum temperature at which a distinct carbon dioxide ocean can exist. Fluids whose temperatures are above the critical point are called *supercritical*, and partake of some features of both liquids and gases. The critical point for molecular hydrogen is just 33K, which is why Jupiter and Saturn do not form hydrogen oceans even when the gas molecules are squeezed together in the high-pressure (but warm) interior. Even silicate 'rock' can fail to form a distinct surface in some conditions. Under high temperature and pressure conditions, molten silicate can actually dissolve into an overlying hydrogen atmosphere, and vice versa, leading to a smooth, mushy transition

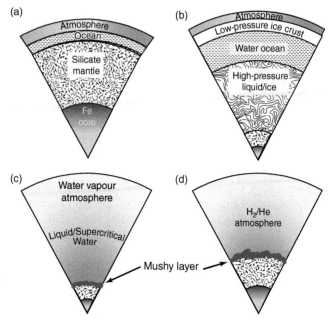

10. A few representative planetary structures. Each can exist in a number of variants, such as differences in the proportions of the various layers. (a) represents a rocky planet with a thin atmosphere and ocean. (b) and (c) represent planets with a dominant water composition. (b) is a cold version with a substantial icy component, whereas (c) is a warmer version in which water appears only in gaseous and fluid form; depending on temperature, there may or may not be a density discontinuity separating the fluid from the gas layer. (d) is a hydrogen world. Massive hydrogen worlds may not have any solid central region at all since the original solid nucleus that attracted hydrogen could have entirely dissolved in the hot gas near the centre of the planet. Where a rocky silicate mantle persists, if the temperature and pressure at the base of the gas/fluid layer is high enough, the interface may take the form of a mushy layer in which the silicate and gas/fluid intermingle, as indicated in the figure.

layer rather than a sharp boundary between a rocky mantle and the overlying atmosphere.

There are many different kinds of planets that can be assembled from the materials described in this chapter. A small sample of the variety is summarized in Figure 10. We'll see how these manifest themselves amongst the exoplanets, and in what ranges of sizes, in Chapter 5. In the Solar System worlds with massive hydrogen or water/ice envelopes appear only as gas or ice giants, but we'll see in Chapter 5 that such planets can also occur in much smaller versions.

Chapter 5
A grand tour of exoplanets

More than two millennia ago, the Greek astronomer Aristarchus of Samos conceived the idea that the stars are suns like our own. He even inferred that they are very distant, on the grounds that their positions didn't show a measurable shift as the Earth proceeds in its orbit. The shift, known as parallax, could not be detected by instruments available to Aristarchus, so he couldn't know precisely how far away the stars were save that they were much farther away than the Sun and the planets. Modern measurements of parallax allow accurate determination of stellar distances, with the spaceborne Gaia mission having pushed the range of the method out to 30,000 light years from Earth for sufficiently bright stars—nearly a quarter of the diameter of our galaxy. The line of thinking begun by Aristarchus was embellished in the 16th century by the Dominican friar Giordano Bruno, who postulated that the stars were not only suns like our own but had their own planets, populated by other-worldly beings. Bruno is considered to be a kind of unofficial patron saint by much of the exoplanet community, as he was widely believed to have been martyred for his beliefs about what we now call exoplanets. He was indeed burned at the stake by the Roman Inquisition, but in fact there is little about exoplanets (even inhabited ones) that is in conflict with Catholic doctrine. Your author even had the pleasure of attending a Vatican meeting on astrobiology, which amongst ourselves we referred to as 'E.T. Phones Rome'. Bruno was an all-round heretic, and had numerous beliefs which clashed much more fatally with Catholic dogma. Be that as it may, it was not until the last few decades that we finally began to get a glimpse of these worlds Bruno imagined. In this chapter, we will learn something about them.

On the naming of exoplanets

There are far too many known exoplanets already for many of them to get familiar names such as the Solar System planets have, so they are identified by a two-part naming convention. The first part of the name is the star they orbit, and the second part is a lower case letter—it begins with the letter b—indicating the order in which the planet associated with that star was discovered. Planets are named in order of discovery, rather than in order of distance from the star, since it is often the case that an outer planet is discovered before an inner planet, and it would be unwieldy to have to rename all the other planets every time a closer-in one is discovered. Most stars do not have names either, so they are referred to either by a catalogue number (e.g. GJ436 for star 436 in the Gliese catalogue of low mass stars) or by the name of the astronomical project that first discovered a planet about the star, followed by the sequence number in the catalogue of the project. WASP-12 is a star examined by the WASP array of ground-based telescopes, and the ultra-hot gas giant Wasp-12b is the first planet discovered orbiting it. Trappist-1 is the first star examined by the Trappist ground-based project to yield a planet, and Trappist-1c is the second planet discovered about that star. Kepler-1000b is the first planet found about star 1,000 amongst the planet-bearing stars found by the Kepler space telescope. A few stars have common names, such as Vega or Polaris or Proxima Centauri (the closest star to the Solar System). Others are named according to the constellation in which they appear, e.g 55-Cancri, which is visually located in the constellation of Cancer the Crab.

Stars, like T. S. Eliot's cats, can have several names, as they often appear in multiple catalogues, and the few stars that have common names also generally appear in one or more catalogues.

How are exoplanets found and characterized?

There are two main ways that exoplanets are found (Figure 11). The first is the radial velocity (RV) technique. This technique makes use of the fact that the gravitational tug of a planet orbiting its star causes the star to wobble a bit about its centre of mass, as shown in Figure 11. Because of the wobble, the star alternates between moving towards and away from a telescope on Earth peering at the star. When the star is moving towards the observer,

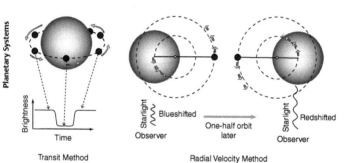

Transit Method   Radial Velocity Method

11. **The two main methods for detecting and characterizing exoplanets.** For the transit method, the planetary system is depicted as seen by the observer, and the graph shows the brightness of the system vs. time, as seen by a telescope in the Solar System. For the radial velocity method, the system is depicted as it would appear looking down from above the plane of the orbit, but the observer would be viewing the system from a more edge-on orientation as in the transit method (although the system need not actually be transiting as seen from the position of the observer). The white dot in the sketch of the RV method indicates the centre of mass of the system. The size of the planet relative to the star, and the magnitude of the displacement of the star in the RV method, have been exaggerated for the sake of clarity. For both of these methods, the observation itself consists of a single time-varying point of light from the planetary system.

its light is shifted slightly towards shorter (bluer) wavelengths, and when moving away, it is shifted towards longer (redder) wavelengths. This is the same Doppler shift phenomenon we have encountered previously. Because planets are much less massive than stars, the velocity of the stellar wobble is very small—hardly more than the speed of a brisk trot—but by the 1990s astronomical instrumentation had become sensitive enough to detect the tiny Doppler shift. The first announcement of an exoplanet detected by the RV method was a planet orbiting the star Gamma Cephei A in a 2.5 Earth-year orbit. Announcement of the discovery was made in 1989 but doubts were raised about the detection owing to the misclassification of the nature of the star; it was not confirmed that the detection really was a planet until 2002.

In 1992, a planet was found to be orbiting the pulsating neutron star PSR B1257+12; this was the first confirmed detection of an exoplanet, but the discovery was made by examination of timing variations of the pulsar, rather than the more broadly applicable RV technique. The first confirmed exoplanet orbiting a main sequence star, 51 Pegasi, was discovered in 1995 using the RV technique, and many more followed shortly thereafter. The watershed moment of 1995 was due not so much to an improvement in instrumentation as a change in the search strategy. At this time, several groups began adjusting their algorithms to look for massive gas giant planets with short-period orbits very close to their host stars, because those kinds of planets would be most easily detected by the RV method. At the time, there was little reason to suspect that such planets would exist. The strategy was rather like looking for one's lost keys at night under a lamp-post, because that's where they'd be most easily found, even if one was pretty sure the keys were lost somewhere in the dark alleyway. Nonetheless, in one of those surprises the Universe so often serves up to theorists, it turned out that such planets—now called hot Jupiters—do exist. The planet 51 Pegasi b, discovered in 1995, has a mass about half that of Jupiter and orbits its star with a period of 4.2 Earth days. It is indeed hot, with an upper atmosphere temperature of 1,284K, as compared to a chilly 84K for our own cold Jupiter. The discovery of 51 Pegasi b eventually netted a Nobel prize for its discoverers.

The RV technique directly yields the period of the orbit, via the period of the stellar wobble. Newton's laws of gravitation imply that the period of the orbit increases with the planet's mean distance from the star. Given an estimate of the star's mass the relation can be made quantitative, yielding an estimate of the planet's orbital distance. From the amplitude of the stellar wobble, one gets the planet's mass, or more precisely the planet's minimum mass since the same mass planet causes less wobble in the direction of the observer if the orbit is viewed obliquely than if it is viewed edge-on.

The other main discovery technique is the transit method. If the orbit of the planet is lined up just right relative to an observer in the Solar System, the planet blocks a bit of starlight when it passes in front of the star. This tiny reduction in brightness also became detectable with astronomical instrumentation in the 1990s. Given that planets are small compared to stars, and their orbits need to be lined up just-so to yield a transit, one might think that transiting systems are rare. Indeed they are—but it turns out that there are so many planet-bearing stars close enough to the Solar System for transits to be observable that the technique has yielded a good crop of planets. In fact, the transit method is responsible for the preponderance of planets that have been detected to date. The transit method also benefits from the possibility for a single telescope to monitor many thousands of stars for the subtle variations in brightness that indicate the presence of a planet. The Kepler space telescope, dedicated to finding planets by the transit method, opened the floodgates and quickly added 2,000 new exoplanets to the roster. The TESS space telescope is poised to add another 20,000 in the near future. The transit method yields the period of the orbit if one observes two or more transits, from which the orbital distance can be inferred. Because the amount of starlight blocked increases with the size of the planet, the transit method yields a measurement of the radius of the planet, based on knowledge of the radius of the star. It does not determine the mass of the planet, but if the star hosts two or more transiting planets (a not uncommon situation) then the gravitational tugging of one planet on another causes subtle variations in the timing of the transits, from which planetary mass can be inferred. In the most fortunate of circumstances, the transiting system can also be observed by the RV method, in which case one knows the mass of the planet precisely since one knows the system is being observed edge-on. With both the mass and the radius, one gets the mean density of the planet, giving some indication of what the planet is made of.

Currently, most observations of exoplanets proceed from observations of the combined light from the star and planet, which appears as a single dot on (more or less) a single pixel of the camera attached to the telescope. Such observations are sometimes called 'single pixel astronomy', and hark back to Carl Sagan's musing about what could be learned about Earth and its inhabitants from the pale blue dot that it would cast on a pixel. There are emerging techniques that allow the starlight to be blocked, so that the single pixel in question corresponds to light reflected from or emitted by the planet alone. For now, techniques using combined star/planet light reign.

Single pixel astronomy can provide a wealth of information about a planet's atmosphere—even though that one pixel blurs together the light from the star and planet. Consider the size of the planet as revealed by transit depth. If the planet is bare rock, it will appear to be the same size at all wavelengths, since all wavelengths are completely blocked by the planet. Now suppose the planet has an atmosphere that is more transparent at some wavelengths than others. Atmospheres do not have a sharp boundary at the top separating them from outer space, but to keep things simple let's imagine an atmosphere of depth 500km, lying above a rocky surface at radius 5,000km from the planet's centre. Suppose this mythical atmosphere is completely transparent to green light, 50% transparent to blue light, and completely opaque to red light. Then, looking at it with three different filters on our telescope, in transit the planet would appear to have a radius of 5,000km when viewed in green light, 5,500km when viewed in red light, and about 5,250km when viewed in blue light. Real atmospheres have densities that peter out to zero gradually with increasing distance from the planet's centre, but the effect is essentially the same as in our simple example. Looking at how the size of a planet, as revealed by transit depth, varies with wavelength allows us to measure the absorption spectrum of an exoplanet atmosphere hundreds of light years away, just as if we could grab a bit of that atmosphere, take it back to our lab, and squirt it into our handy spectrometer. Every gas has a characteristic spectral fingerprint. For example, water vapour absorbs infrared light with a wavelength of 4 microns very strongly, but is relatively transparent to infrared light with a wavelength of 5.7 microns, so if a planet looks bigger at 4 microns but smaller at 5.7 microns, we could conclude that there is water vapour in the atmosphere. This technique is called *transit depth spectroscopy* and has become the workhorse of atmospheric characterization.

The temperature of a planet can be determined from single pixel astronomy by inferring how much infrared the planet emits, since hotter objects emit infrared at a greater rate, in accord with the blackbody law. The planetary emission for a transiting planet can be inferred by looking at the dip in infrared in the combined star/planet light which occurs when the planet goes behind its star (the *secondary eclipse*). In this technique, instead of looking at transit depth, which involves blocking of starlight by the planet, astronomers look at eclipse depth, which involves blocking of the planetary emission by the star. Eclipse depth provides a measure of the dayside temperature of the planet. If one is so fortunate as to get enough telescope time to observe one or more full orbits of the planet, the resulting time series—called a *phase curve*—provides information about the geographical variations in temperature, since one gets to look at the planet from all different angles as it proceeds through its orbit. Because gases emit and absorb infrared better at some wavelengths than others, the spectrum of a planet's emission provides an additional window into atmospheric composition.

The population of exoplanets

So what kinds of planets have we found out there? Planets can be characterized by their instellation—the rate at which energy from their stars illuminates them. The instellation depends on both the power output (luminosity) of the star and the distance of the planet's orbit from the star. A planet orbiting a dim M star would need to be closer in than a planet orbiting a brighter G star in order to receive the same instellation. If instellation is measured in units of Earth's average instellation, a planet with unit instellation would have a surface temperature similar to Earth's, if given an Earthlike atmosphere and surface conditions—a big 'if'. A planet with an instellation of 10 Earth units would be a very hot place, and one with an instellation of a tenth of an Earth unit would be a very cold place. We'll learn more about the temperature of planets in Chapter 6.

Planets can also be categorized by size or mass, depending on which measurement is available. A planet of the size or mass comparable to or greater than Jupiter is almost certainly a hydrogen-dominated gas giant, whereas a much smaller or less massive planet has a better possibility of having a predominantly rocky or icy composition. Information about instellation is put together with planetary size or mass in Figure 12. This is not a random sample of the planets in our vicinity, as some kinds of planets are easier to detect than others. For example, very large or massive planets are easier to detect than Earth-sized planets. Planets orbiting close to their stars are easier to detect than those orbiting further out.

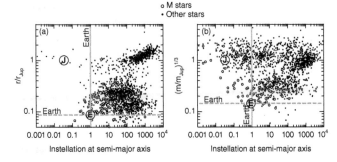

12. Scatter plot of planet size (a) and planet mass (b) vs. installation. Installation is given relative to Earth's current installation, while radius and mass are in units of Jupiter's values. For the mass plot, the vertical axis is expressed in terms of the cube root of the mass, since for a given mean density of the planet this quantity is proportional to the radius. Thus, a planet with Jupiter's mean density would show up plotted with the same value on both the radius and mass axes. Planets orbiting M stars are plotted as open circles. On each plot, the position of Jupiter is marked by a J and of Earth by E.

These plots show a great diversity of planets unlike any seen in our Solar System. There are the Hot Jupiters—gas giants with mass or radii comparable to or larger than Jupiter but receiving up to ten thousand times as much installation as Earth. Jupiter itself receives just 3.7% of Earth's installation. Hot Jupiters were the first kind of exoplanets discovered, but prior to their discovery there was no reason in then-extant planetary formation theory to believe such things should exist. Gas giants could not form in the hot regions close to the star, but it was thought that they could also not migrate inwards from the farther, cooler reaches of the disk before the disk dissipated, and gas drag with it. Their existence was an ugly fact, which forced a major revision of the theory of planet formation and migration. Gas giants come in the full range of temperatures, ranging from cold Jupiters receiving about the same installation as our own Jupiter through temperate Jupiters receiving the same installation as Earth, through to Hot Jupiters, and everything in between. Large moons orbiting temperate

Jupiters could have a rocky or icy composition, and could provide habitable surfaces, though no exo-moons have yet been discovered.

In the Solar System there is a clear dichotomy between small, rocky planets (Earth, Mars, Venus, and Mercury) and the ice giants Uranus and Neptune; no such gap exists amongst the exoplanets. There is a continuum of masses and sizes ranging from below that of Earth to Neptune-sized (3.9 Earth radii or .34 Jupiter radii). The intermediate-sized planets in this range may be rocky 'Super-Earths' or gassy/icy 'mini Neptunes', and discriminating between the two is an active subject of research. These planets are prevalent even in orbits with high instellation that would prevent the formation of ice in the protoplanetary disk. Like the Hot Jupiters, they may have formed further out and migrated inwards, but the mechanism of their formation is still largely unresolved. Although there is no clear gap between Earth and Neptune radii, there is a radius gap between Neptunes and gas giants, which cries out for an explanation. There is also a notable paucity of Neptune-sized objects in the high-instellation region, which cannot be attributed to observational bias. This is called the Neptune Desert. Why do we have Hot Jupiters but not Hot Neptunes? Merely warm Neptunes, on the other hand, are quite abundant.

Further information about the possible composition of the exoplanets can be obtained from the subset for which both mass and radius information is available, shown in Figure 13. Here, the cube root of mass is plotted against the radius. For a given radius, high-density planets, such as those dominated by rock or iron, will have a higher mass than low-density icy or hydrogen-rich ones. As a guide, lines of constant density are plotted on the graph. At low pressures (e.g. on the surface of Earth), iron has a density of 7,800kg/m^3 and silicate rock comes in at around 3,000kg/m^3. The Earth's core makes up about a third of the mass of the Earth, and based on surface pressure densities the mean density of the Earth would be 3,800kg/m^3. Interpretation of density data, however,

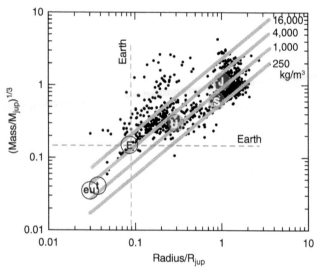

13. Scatter plot of mass (vertical axis) vs. radius (horizontal axis). Grey lines are lines of constant density. This plot provides some indication of the composition of the planets indicated by dots on the graph. Generally speaking, a planet with a density of 4,000kg/m³ or more can be expected to be rocky, with greater densities indicating a massive iron core. Densities between 1,000 and 2,000 kg/m³ indicate a dominant icy composition, and lower densities require hydrogen/helium-dominated compositions such as that of Jupiter or Saturn. Because of the compressibility of hydrogen, some gas giant planets can have high densities without being icy or rocky, as described in the text. The positions of Earth, Uranus, Saturn, and Jupiter on the mass-radius plot are indicated by E, U, S, and J. Those of the icy moons Europa and Titan are indicated by eu and t.

must contend with the compression of materials at the enormous pressures of planetary interiors. More massive planets have higher interior pressures and squeeze their materials into smaller volumes, and moreover some materials (notably hydrogen and ices) are more compressible than others. Earth's actual mean density of 5,500kg/m³ is considerably higher than the value we estimated neglecting compressibility. An Earth-sized planet with the density of Earth is likely to have a rock/iron composition, whereas one with somewhat higher density would be more iron-dominated and one with somewhat lower density would be more rock-dominated. For larger planets, up to, say, 2 Earth radii, the characteristic densities would need to be scaled up a bit because of greater compression.

From this crude analysis (which can be made more precise using actual compressibility properties of various planetary materials) we can see that while a good number of planets in the 1–2 Earth radius range fall in the density band corresponding to a rock/iron composition, quite a few have surprisingly low densities, as low as $1,000kg/m^3$ or even lower. Even at Earth surface pressure, liquid water has a density of $1,000kg/m^3$, with ice being slightly less; under the high pressures of a planet with 2 Earth radii, the mean density of a pure water-world roughly doubles. These low-density smallish planets are a completely new category of planet, unlike anything seen in the Solar System. Even the ones with density as high as $4,000kg/m^3$ are too light to be made of pure silicate rock, owing to the effects of compressibility. These low-density planets must have a substantial low molecular mass envelope, which may take the form of gas, liquid, or ice, in some combination. Hydrogen is especially good at puffing up the size of a planet as measured by transit depth, and in most cases the density can be matched by the addition of a moderately massive hydrogen envelope to a rocky core. However, except for the lowest density planets, the density can equally well be matched by assuming no rocky or iron component at all, and making the planet entirely out of a low molecular mass substance such as water, methane, ammonia, or carbon dioxide (which would take the form of high-pressure fluid and ice in the interior, as in the conventional model of Uranus and Neptune). For example, the M star planet GJ1214b has a radius 2.85 times that of Earth, but a mass of only 6.26 Earth masses, giving it a mean density of just over a quarter of Earth's, or $1,500kg/m^3$. Allowing for the compressibility of water, it could be a pure water-world, but alternatively it could be composed of a rocky core surrounded by a thick hydrogen atmosphere. This class of planets is so newly arrived on the scene that planetary scientists are still scratching their heads about what to call them. Prosaically, they've been called 'Sub-Neptunes', but terms like neptines, neptinis (i.e. a teeny Neptune, which would also be a rather good name for a cocktail), neptinos, and gas dwarfs (in contrast to gas giants) have been suggested. Whatever they are called, the problem of accounting for the formation of smallish planets with a very high volatile inventory poses a great and largely unsolved problem in planetary science.

But what are we to make of the high densities of some of the planets in the gas giant size range? Are these in fact 'rock giants'? Actually, these planets achieve their high densities through compression of hydrogen and helium, rather than through incorporation of high molecular mass substances. For the gas giant mass range, the interior pressures are so high, and hydrogen/helium mixtures are so compressible, that past a point adding more mass just crams more gas into the same volume, and the radius stops increasing. This is known as 'hydrogen degeneracy'. Saturn, with a third of the mass of Jupiter, has a density of $687kg/m^3$, in contrast to Jupiter's value of $1,330kg/m^3$. If one adds enough mass rapidly enough the interiors of gas giants get so hot that the radius can start increasing again, simply because hot gas is less dense than cooler gas. In fact some of the low-density gas giants are 'inflated hot Jupiters', which appear to require higher internal temperatures than standard planetary formation theory can account for. At some point, if one adds enough hydrogen, one ignites fusion and builds a star with very high interior temperatures rather than a planet, which is why stars can be considerably larger than Jupiter.

Tide-locked planets

In the Solar System, the shortest orbital period, or year, is that of Mercury, at 88 Earth days. Many exoplanets have much shorter orbital periods, down to as low as a few hours. This includes all the high installation planets, such as Hot Jupiters, but it also includes planets with Earthlike installation orbiting dim, red dwarf M stars. The nearest star to the Solar System is the red dwarf Proxima Centauri, and a planet with a mass somewhat greater than Earth—Proxima 1b—has been found orbiting it in an eleven day orbit where it receives 72% of the Earth's installation.

M stars often form compact planetary systems, with several planets in short-period orbits. This includes the remarkable Trappist 1 system, with seven planets squeezed into orbits that would fit within a fifth of the orbital distance of Mercury. Their periods range from 1.5 to eighteen Earth days. The planets are so close together that at the time of closest approach Trappist 1e as seen in the night sky of Trappist 1d would look considerably bigger than Earth's Moon appears to us (Figure 14). Interplanetary travel

14. Trappist 1e and 1f as seen in the night sky of Trappist 1d, viewed at a time when both planets appear near the horizon. Big Ben and the Houses of Parliament are placed in the foreground 2km from the observer, to provide a sense of scale. Trappist 1e and 1f are shown near conjunction (i.e. when the two are close together in the sky) at a time of their closest approach to Trappist 1d.

in such a system would be hardly more difficult than the voyage from the Earth to the Moon.

For planets whose orbits lie close to their stars, the gravitational force on the part of the planet closest to the star is considerably greater than that on the distant side. This stretches the planet out into an ellipsoid, with a bulge on each side of the planet in similar fashion to Earth's ocean tides. The tide can manifest as deformation of solid rock or a thick gaseous atmosphere, depending on the composition of the planet. Planets are born spinning, because they inherit the angular momentum of the parent disk, but if a close-orbit planet is spinning too rapidly, the rotation carries the tidal bulge away from its original position, requiring the tidal forces to do work on the planet to maintain the bulge. Dissipation from the continual flexure of rock and fluid turns the planet's rotational energy into heat, and forces on the distorted planet act to reduce its spin rate. For a planet in a circular orbit, the process ends when the planet becomes 'tide-locked' and always presents the same face to its star just as our Moon, which is tide-locked to Earth, always presents the same face to Earth. A tide-locked planet is still spinning about its axis. An observer would still see the constellations rotate across the celestial sphere over a period called the sidereal day. On a tide-locked planet, the sidereal day is equal in length to the planet's orbital period (i.e. its year), so that the turning of the planet keeps pace with its progress in its orbit, always keeping the same face presented to its star. All planets would eventually spin down to a tidally locked state given enough time. For planets in distant orbits, such as Earth, the process takes a very long time and many planets would be destroyed by events at the end of their star's main sequence lifetime before a tide-locked state is approached. For planets in orbits close to their stars, the time taken to become tide-locked is very short, from as little as a few decades to a few million years. Since orbital period increases with distance from the star, planets in short-period orbits are more likely to be tide-locked than planets in long-period orbits. Rocky planets with periods of roughly thirty Earth days or less are likely to be tide-locked, though the precise value of the orbital period where tide-locking becomes likely depends somewhat on the mass of the star.

A tide-locked planet has a perpetual dayside and a perpetual nightside. If you remained in one place on such a planet, your sun would never rise or set. It would incessantly maintain the same position in the sky, and denizens of the nightside would never see their sun at all (an experience Arctic peoples on Earth have some familiarity with, for several months a year). For a planet with an atmosphere, the air currents stirred up by the stark contrast between dayside and nightside heating would create meteorological phenomena different from anything we know in the Solar System. The same underlying physics that applies to the Earth's atmosphere applies to tide-locked atmospheres as well, but the novel way these principles play out demands a virtual reinvention of the subject of dynamic meteorology.

Music of the spheres, dance of the planets

In the 6th century BC, the Greek mathematician Pythagoras discovered that the pitch of a string of a lyre is inversely proportional to its length: a string 1/2m long sounds exactly twice as high as a string 1m long—what we would now call an octave.

Although Pythagoras had no way of measuring the actual frequency of sound, in modern terms the frequency of the shorter string is twice that of the frequency of the longer one. The period of the oscillation is the time it takes for the plucked string to repeat its position, and is equal to the reciprocal of the frequency. All other things being equal, the period of sound made by a 1m string is twice the period of the sound made by a 1/2m string. Such notes sound harmonious when played together. Pythagoras found that other intervals in the proportion of small integers also sound harmonious, of which the most magical is the fifth, which has frequencies in the ratio of 3:2 of the higher note relative to the lower note—like playing the notes C and G. Pythagoras constructed a musical scale around a circle of fifths, which includes other especially harmonious intervals like the fourth, in a ratio of 4:3. The Pythagorean scale, with various modern adjustments, forms the basis of all Western music.

Pythagoras was so taken by the beauty of the laws of harmony that he posited that the whole Universe was pervaded by an inaudible cosmic hum caused by the motion of the celestial bodies. The 'pitch' of a planet would be the planet's orbital frequency, which is the reciprocal of the planet's orbital period. Planets in long-period orbits distant from their star sing with the lowest pitches, while planets in close orbits which whizz around quickly sing with the highest pitches. There was scant evidence of celestial harmony at the time of Pythagoras, or even that of Kepler. Mercury, Venus, Earth, Mars, Jupiter, and Saturn orbit with the discordant frequencies—measured relative to the orbital frequency of Saturn—of 120.36, 47.07, 29.02, 15.43, 2.42, 1. Later, it was found that the Jovian moons Io, Europa, and Ganymede orbit in Pythagorean octaves with frequencies 4, 2, 1 relative to Ganymede's frequency. Much later it was found that Neptune and Pluto orbit in a 3:2 Pythagorean fifth. Planets or satellites whose orbital frequencies are in ratios of small integers are said to lie in a resonant chain, by analogy with resonant strings of musical instruments which are excited by harmonics of a plucked or bowed string.

But Pythagoras would have loved the Trappist 1 system. Let's take as our unit of time *twice* the orbital period of the outermost planet Trappist 1h. This makes the orbital frequencies come out as nice integers, as Pythagoras would have liked. The seven planets Trappist 1 b–h then orbit with frequencies close to 24(b), 15(c), 9(d), 6(e), 4(f), 3(g), 2(h). Trappist 1h sings with the lowest pitch, and Trappist 1g sings in harmony at a pitch a Pythagorean fifth higher. Trappist 1f makes a perfect Pythagorean fourth with 1g (an octave above 1h), and Trappist 1e harmonizes at a Pythagorean fifth higher than 1f (an octave above 1g). Trappist 1d is a Pythagorean fifth above 1e. Trappist 1c is slightly out of tune relative to the Pythagorean system, but is close to being a Pythagorean minor seventh ($2^4/3^2$) above 1d and Trappist 1b is close to being a Pythagorean fifth above 1c. Many other harmonies can be spotted within this beautiful system, and indeed one can compose interesting music just with the notes present in it. Musically speaking, the architecture of the Trappist 1 system corresponds to a seven-string lyre, with the shortest string being 4 1/6 (i.e. 100/24) cm long and the longest being 50cm.

A geometric sense of the rhythm of the orbits can be gleaned from Figure 15. This figure shows a sequence of snapshots of the

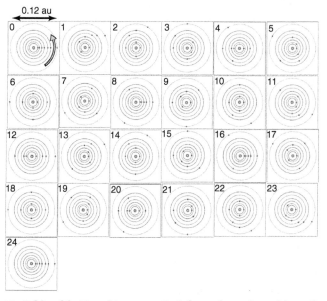

15. Orbits of the Trappist 1 system. Each frame shows the positions of the planets after the indicated number of orbits of the innermost planet, Trappist 1b. Positions are shown as if the planets are initially lined up, though the actual repeating configuration is not colinear.

positions of the planets, taken at an interval of one Trappist 1b year (which is only 1.51 Earth days long). To make it easier to spot recurring positions, the patterns are shown as if the planets are initially all placed on a line, although this doesn't correspond to the actual phase of the orbits. The recurrences of various patterns correspond to the resonances in the system, and the entire sequence repeats at intervals of 24 Trappist 1b orbits.

Trappist 1 offers the longest resonant chain known so far, but shorter chains have been found in other systems. For example, the system K2 138 has five planets in a nearly unbroken 3:2 chain of fifths. Resonant chains are far from ubiquitous, but are not rare either.

The gravity of a planetary system is by far dominated by the gravity of its star, but the subtle gravitational tugging exerted by one planet on another can render some orbital resonances very stable against disruption, although the details of which resonances are stable and which are unstable depend on the masses of the planets involved and how closely they approach each other. The dissipation that returns a perturbed system to its stable resonant configuration can be provided by gas friction in the protoplanetary disk and tidal dissipation within the planets, of which the latter persists after the disk dissipates. As planets drift inwards through the action of gas friction in the disk, they can lock into a stable resonance and remain in that condition as they drift, and are then frozen in the resonant state when the disk dissipates.

Migration within the disk does not generally lead to resonant configurations, and planets can move outwards as well as inwards. The 'Grand Tack' model of Solar System evolution argues that Jupiter and Saturn drifted inwards because of drag against the gas of the disk, with Saturn migrating faster than Jupiter. As Jupiter moved inwards, it gobbled up material that could have made Super-Earths or Neptinis, and rocky material that could have made Mars a larger planet. As Saturn caught up with Jupiter, the two became locked in a 3:2 orbital resonance (a Pythagorean fifth again!). If Jupiter had continued its inward migration and become a Hot Jupiter, we wouldn't be here to talk about such things, as its inward migration would have severely disrupted Earth's orbit. Fortunately, when it reached a distance 1.5au from the Sun, the joint effect of Jupiter and Saturn on gas flow changed the force balance so that the gas began to *accelerate* the planets, leading to outward migration. As the disk dissipated, the outward migration ceased, but at different rates for Jupiter and Saturn. This broke the resonance, leaving the planets in the unharmonious 29:12 period ratio we see today. The Grand Tack Hypothesis is far from proven, but it provides a plausible account of the character of the inner planets and asteroid belt.

Migration happens most swiftly during the brief period before the protoplanetary disk dissipates, but can continue much more slowly afterwards through the effect of tidal dissipation and gravitational interactions between the planets. Systems with fairly circular orbits can remain stable for billions of years, as seems to have been the case for our Solar System.

The very word 'planet' comes from the Greek word for 'wanderer', and the planets were so called because planets wandered among the (relatively) fixed stars as seen in Earth's night sky. We now know that the term is even more fitting than ancient Greek astronomers could have imagined, as the wandering of planets crafts the architecture of planetary systems, and profoundly affects their habitability.

Chapter 6
Planetary climate and habitability

In Chapter 5, we used instellation as a general index of how hot a planet would be relative to Earth. In this chapter, we make a deeper dive into the subject of planetary climate, and particularly the way an atmosphere affects temperature and the suitability of a planet's surface as a home for life as we know it.

The temperature of planets

A planet's temperature is not characterized by just a single number. The entire temperature profile, from the planet's centre to the tenuous outer reaches of its atmosphere (if it has one) is of importance. The temperature of Earth's core is something over 6,000K whereas the mean temperature of our continental and ocean surface is around 287K and the coldest point of our atmosphere is close to 190K. The surface temperature is especially meaningful to us because that's where we live, as well as most life with which we share the planet. Ocean life is abundant, but our ocean is a thin layer compared to Earth's radius, and the deepest ocean temperature is only around 10K colder than the mean surface temperature in the modern climate, so it can be considered to be in the same range as surface temperature.

As a first step towards understanding planetary temperature, let's consider an airless rocky planet. Heat conducts very slowly through solid rock—it is a very good insulator—so the surface temperature depends little on the interior temperature of the planet. Horizontal heat conduction within rock is also very weak, and so it carries very little heat from one point to another over the planet's surface. Earth's Moon is a good realization of this ideal situation. (Ideal for physicists at least, if not for inhabitants.) In this case, the temperature at any point at the planet's surface is determined by energy balance, and specifically the requirement that the rate at which energy is absorbed from the star equals the rate at which energy is lost to space. There is only one way that planets are able to shed a significant amount of energy, and that is by electromagnetic radiation. We encounter again our old friend, blackbody radiation. The rate of emission of blackbody radiation increases like the fourth power of temperature. At the substellar point, where the star is directly overhead, the rate of energy absorption from the star is the instellation, minus the proportion (called *albedo*) reflected back to space. For a given albedo, the temperature of the substellar point on a bare-rock planet then goes like the fourth root of instellation. As a point of reference, for Earth's instellation, the substellar temperature for a zero albedo (perfectly absorbing) surface is 394K—well over the sea-level boiling point of water. This is quite close to the maximum temperature of the Moon, which shares Earth's instellation. The moon may look silvery, but it is only bright in comparison to the dark night sky. Objectively speaking, it is about as dark as coal and so approaches the zero-albedo idealization. The basic energy balance that determines planetary temperature is illustrated in Figure 16.

Energy balance also tells us how temperature varies over the surface of a bare-rock planet. Imagine that you are suited up in your best spacesuit and going for a stroll from the substellar point towards the terminator, which is the line separating day (where

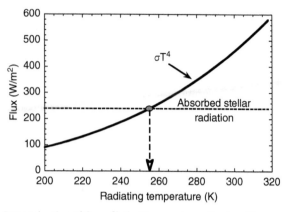

16. Determination of the radiating temperature of a planet by energy balance. For a planet whose dominant energy source is absorption of light from its star, the radiating temperature is determined by the requirement that the planet must emit energy at the same rate at which it is absorbed. Since the rate at which energy is radiated is proportional to the fourth power of temperature, the radiating temperature goes like the fourth root of the absorbed stellar flux. For a bare-rock planet, or a planet whose atmosphere is transparent to infrared radiation, the radiating temperature is the temperature of the ground itself. For a planet whose atmosphere transports little heat, the energy balance can be applied at each point of the planet's surface individually. For a planet whose atmosphere transports heat very efficiently, the radiating temperature is geographically uniform, and the appropriate absorbed stellar flux to use in the balance is $\frac{1}{4}(1 - \alpha)I$, where I is the instellation and α is the proportion of starlight reflected back to space.

the star is above the horizon) from night (where the star is below the horizon). To see how the incident stellar radiation is spread over the surface of the planet, imagine a square of cardboard 1m on a side, held up so that its perpendicular points directly towards the centre of the star. The energy hitting this square is by definition the instellation of the planet, but where the star is lower on the horizon, the square casts a longer shadow. The shadow cast represents the area over which the instellation would be spread if the cardboard were removed. Thus, as one approaches the terminator, the instellation is spread over greater and greater areas of the planet's surface, leading to fewer Watts per square metre of the planet's surface being received; once the star falls below the horizon, the energy per square metre of surface received from the star falls to zero. For those who remember their trigonometry, the stellar energy per square metre of surface at any given point is the instellation times the cosine of the zenith angle, where the zenith angle is the angle between the line pointing to the star and the local normal to the spherical surface—basically a stick stuck into the surface pointing towards the planet's centre.

Each square metre of a planet's surface emits infrared at a rate proportional to the fourth power of its temperature, so in equilibrium the temperature is proportional to the fourth root of the stellar energy received per square metre of surface, which falls to zero as the terminator is approached. Now, the fourth-root is a rather sluggishly varying function, so that the temperature tails off rather gradually until the terminator is approached quite closely. For Earth's instellation, by the time you had trekked halfway towards the terminator (rather like walking from the North Pole to Bordeaux) the temperature would have dropped off from the torrid 394K substellar value to the only marginally less torrid 361K—still hot enough to make a passable cup of tea by just setting the kettle on the ground. At 2/3 of the way, the temperature is 331K and you still can't turn off your suit's air conditioner. From there on, though, temperature falls off rapidly, and by the time you are 4/5 of the way and the star is low on the horizon, it is a chilly 254K. None of this is dependent on the size of the planet; the temperature pattern would be the same if you were on a giant rocky Super-Earth with twice Earth's radius or on a spherical planetesimal just 100km across, although in the latter case you would not need to walk so far to get from the substellar point to the terminator.

Figure 12 shows a number of smallish and potentially rocky planets which receive thousands of times the instellation of Earth.

Some of them have known densities which confirm their rocky status. Such a planet is 55 Cancri e, orbiting a star in the constellation of Cancer the Crab. Its year is only 3/4 of an Earth day, and it receives just shy of 2,500 times Earth's instellation. The substellar temperature if the planet is bare rock would be 7.07 times the substellar temperature for Earth's instellation, since $7.07^4 \approx 2500$. That comes out to a torrid $7.07 \times 394K$, i.e. 2,785K, which is high enough to melt most kinds of rock. Even 2/3 of the way towards the terminator, the temperature has only fallen to $7.07 \times 331K$, i.e. 2,340K, which is still enough to melt rock. This planet would have a permanent lava ocean extending over most of the dayside of the planet. Many other lava planets are known, including K2-141b and Corot 7b (the first lava planet discovered). These planets may have thin sodium and silicate vapour atmospheres evaporated from the lava ocean, which rain or snow out as the terminator is approached, but some may have retained thicker atmospheres made of gases like carbon monoxide that could fill out a global atmosphere. The situation is still unclear, but instrumentation of the coming decade should begin to provide some answers.

Now let's clothe our bare-rock planet in an atmosphere. Atmospheres can flow, so they are quite good at moving heat from one place to another, and the more massive the atmosphere, the better the heat transport. Our own atmosphere is the reason the nightside of the Earth does not become cold enough to liquefy air—which would happen on the nightside of the Moon if you released a tank of air into the vacuum of the lunar night. Oceans also transport heat, and can be viewed as just an especially dense form of atmosphere. In the limit of infinitely effective heat transport, the unequal distribution of received radiation is evened out, and the whole surface of the planet would have the same temperature. To get that temperature, we balance the energy received from instellation against the energy radiated by the uniform temperature surface. A spherical object casts a circular shadow, so the area of the stellar beam intercepted is the cross-sectional area of the sphere. However, the body radiates from its entire surface area, which is four times as great as the cross-section area. Reaching back into our basic geometry, the area of a circle of radius r is πr^2, but the surface area of a sphere is $4\pi r^2$. Using the fourth-power law for radiation then we find that the surface temperature for the constant-temperature planet is smaller than the substellar equilibrium temperature by a factor of $1/\sqrt[4]{4}$, i.e. $1/\sqrt{2}$, or about 0.707. That's assuming that the atmosphere was transparent to the infrared radiation escaping from the surface, and that the atmosphere did not reflect back any additional instellation. A pure helium atmosphere on a dry planet would come pretty close to this idealization. Such an atmosphere, given to the moon or a hypothetical dark, bare-rock Earth, would yield a surface temperature of 0.707 × 394K, or 279K. The heat redistributing effect of our atmosphere is the main reason that no point on the surface of the Earth (leaving out volcanoes, geysers, and pizza ovens, which are not in equilibrium with Solar energy) ever becomes as hot as 394K.

The size of a planet doesn't directly affect its temperature, all other things being equal. For a given temperature, a bigger planet radiates more because it has greater surface area, but it also receives more stellar energy, because it has greater cross-sectional area. Both effects are proportional to the square of the planet's radius, and so they cancel out.

Real atmospheres are never completely transparent to the infrared radiation which seeks to escape to space. Gases which are good absorbers of infrared radiation act as planetary insulation. By retarding heat loss they make the layers beneath the atmospheric blanket warmer, for much the same reason that snuggling into a woolly blanket on a cold day makes you feel warmer. This is the *greenhouse effect*, and gases that are good infrared absorbers are called *greenhouse gases*. Almost any gas, including hydrogen, has a strong greenhouse effect if there is enough of it, but some gases are such good infrared absorbers that they are effective greenhouse gases even if present in only trace amounts. Carbon dioxide, which is the source of our own planet's problem with global warming, is an important greenhouse gas for Earth even though only 400 molecules of Earth air per million are carbon dioxide at the time of writing. The number is rising rapidly, owing to burning of fossil fuels; before the Industrial Revolution, the number hovered around 280 molecules per million for thousands of years. Water vapour and methane are also potent greenhouse gases.

A planet must radiate away all the energy it absorbs from its star, plus whatever amount of energy trickles out of the hot interior. The infrared leaving a planet can be characterized by a radiating temperature T_{rad}, which is the temperature a spherical blackbody of the same size as the planet would need to have in order to radiate the same amount of energy as the actual planet. T_{rad} is determined solely by the amount of absorbed instellation and internal heat flux. An atmosphere can reduce T_{rad} by reflecting back some of the instellation, thus decreasing the energy input to the planet, but if the albedo of the planet doesn't change, the atmosphere can't change T_{rad}. The greenhouse effect doesn't 'trap' energy within the planet's atmosphere; if it did, the planet would heat up indefinitely and eventually vaporize. Instead, the greenhouse effect increases the temperature the deep atmosphere or surface has to have in order to lose the same amount of energy it would need to shed if the atmosphere were transparent to infrared. This is the way any good insulator works. Putting insulation in the walls of your house increases the inside temperature by increasing the temperature difference between inside and outside that needs to be reached in order to lose heat at the same rate your furnace is putting it out. Building insulation (or a woolly blanket) works by reducing heat transfer due to air currents, rather than by reducing heat transfer by radiation, but the energy balance principle is the same.

So, T_{rad} is a temperature, but what is it the temperature *of*? The Earth's radiating temperature is 255K, allowing for the amount of instellation clouds and the atmosphere reflect back to space. This is the actual temperature the Earth's surface would have if you could replace the atmosphere with an atmosphere transparent to infrared (such as helium) while keeping the planet's albedo unchanged. The actual mean surface temperature of Earth is about 288K, which is considerably greater than the radiating temperature. However, most of the blackbody radiation from Earth's surface is absorbed by the greenhouse gases in the atmosphere before it escapes to space. The infrared light that escapes to space is emitted from the higher, thinner parts of the atmosphere about 5km above the surface on average. The temperature there is the radiating temperature. The greenhouse effect works by raising the altitude of the level from which infrared radiation escapes to space, while the radiating temperature remains fixed (Figure 17). The concept of the infrared radiating level of a planet is essentially the same as the concept of the photosphere of a star, which is the layer of the star which becomes transparent enough to allow light to escape. The photospheric temperature of a G star like the Sun is around 6,000K although the core temperature is over 10 million Kelvins, and it could be said that the temperature of the stellar core is maintained by a greenhouse effect very similar to the greenhouse warming of a planet's surface, although in the case of a star the energy that escapes to space corresponds to the energy generated by nuclear fusion in the core.

The warming of the surface by the greenhouse effect depends crucially on the fact that the atmospheric temperature decreases with altitude, at least up to the vicinity of the radiating level. Temperature goes down with altitude under circumstances when the heat trying to escape to space originates deep in the atmosphere. The temperature gradient stems from the fact that heat spontaneously flows from hotter places to colder places, at a rate that increases with the temperature gradient. The actual mechanism of the heat transfer could be either electromagnetic radiation (usually infrared) or buoyancy-driven fluid motions.

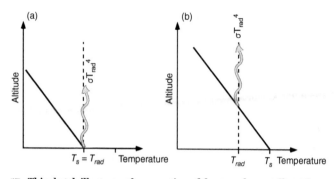

17. This sketch illustrates the operation of the greenhouse effect. The solid line indicates the temperature profile of the atmosphere, which goes down with altitude. For an atmosphere that is transparent to infrared radiation, the radiating temperature is the temperature of the ground itself (Panel a). Adding a greenhouse gas to the atmosphere makes the lower atmosphere opaque to infrared, so that infrared cannot escape from there. It must be transported to higher, less opaque, regions of the atmosphere in order to be radiated away to space (Panel b). The greenhouse gas does not change the radiating temperature of the planet, which is purely determined by the rate at which energy from the planet's star is absorbed by the planet. Instead, the greenhouse case raises the altitude of the radiating layer. The result is a higher surface temperature than was the case without a greenhouse gas, as is evident from extrapolating the temperature profile to the surface to find T_s.

A common circumstance for planets with a distinct oceanic or rocky surface is that much stellar radiation penetrates the atmosphere and is absorbed at the surface, and then has to be transferred to the atmosphere in order for the energy to be carried upwards to the radiating level where it can directly escape to space. On a mostly gassy planet, like a gas or ice giant or Neptini, the temperature gradient in the deep atmosphere can instead be maintained by heat escaping from the hot planetary interior, even if essentially no stellar radiation penetrates to those depths.

The physics of the greenhouse effect that warms Earth's surface is universal and applies to any planet. Because of its highly reflective clouds, Venus has a radiating temperature of only 232K, which is lower than Earth's even though the planet is closer to the Sun than we are. This is indeed the temperature of the upper 1% of the Venusian atmosphere. The thick CO_2 atmosphere of Venus, which has a surface pressure nearly a hundred times that of Earth's, is so opaque to infrared radiation that one has to go to great altitudes in order for any infrared to escape. Because the temperature increases steadily as one goes down from the radiating level to the surface, by the time one reaches the surface the temperature is a scorching 700K.

What does the greenhouse effect do to a planet like a gas or ice giant or Neptini with no distinct surface? The principle of the radiating temperature and radiating level is the same as we discussed for terrestrial-type planets with a distinct surface. In gassy planets, stellar energy is deposited directly in the atmosphere by absorption, rather than being absorbed at a surface and communicated upwards. If stellar energy is deposited deeper in the atmosphere than the radiating level, the temperature in the deposition region will be higher than the temperature of the radiating level, exhibiting greenhouse warming much as if there were a surface there. Since atmospheres are typically more transparent to shortwave stellar radiation than to infrared, the deep atmospheres of gassy planets can be hotter than the radiating temperature. Below the deposition layer, though, if there is no heat that needs to escape, the temperature should remain fixed. More typically, there is sufficient heat in the very deep interior left over from the formation of the planet that the escape of this heat allows the temperature to continue to increase to great depths. To take an example from our own Solar System, the radiating temperature of Jupiter is 126K; based just on the absorbed Solar radiation, the radiating temperature would be only 103K, so there is a considerable amount of heat escaping from the interior. The temperature deeper in the atmosphere at a pressure equal to 1 Earth atmosphere is 167K, and at 7 atmospheres pressure it is an Earthlike 300K. By the time one gets to the 18 atmosphere pressure level the temperature has reached 400K, and we have still gone only a minuscule way towards the centre of the planet which lies at a pressure of over 50 million atmospheres. The radiating temperatures of Hot Jupiters can be thousands of Kelvin, but their interior temperatures can be hotter still, for reasons similar to those operating on our own cold Jupiter. The planetary insulation created by the greenhouse effect keeps the interior of the planets hot, just as a thermos flask keeps your coffee hot. The effect is observable via mass and radius measurements of exoplanets. For a given mass, Hot Jupiters are bigger than our own cold Jupiter not because the outer atmosphere of the planet is hotter than Jupiter, but because the interior is hotter as well. In fact, some Hot Jupiters are so puffy that their inferred high interior temperatures cannot readily be explained by established planetary physics.

Cloudy, with a chance of sapphires

Clouds form when some component of an atmosphere gets cold enough to condense from its gaseous form to a liquid or solid form. Since the condensed phase is denser than the surrounding gas, cloud particles settle out until they reach the planet's surface, or fall to someplace hot enough for them to evaporate back into the gas phase. Cloud particles can aggregate into larger particles that fall faster, forming rapidly falling rain or snow.

Our familiar Earth clouds are made out of condensed water, but on other planets a variety of other gases can form clouds. On Mars, carbon dioxide forms dry-ice clouds. On Jupiter and Saturn there are ammonia as well as water clouds, and on Uranus, Neptune, and Saturn's moon Titan clouds form from methane. On hot planets elsewhere in the Universe clouds form from substances we normally think of as rocks or minerals on Earth. On the Neptini GJ1214b, salt clouds made of potassium chloride or zinc sulfide can form, and on Hot Jupiters clouds form from a variety of silicate minerals. They can even form from metal oxides such as Al_2O_3, which in its gemstone form is known as sapphire (although it more commonly occurs in the form of corundum, which is the stuff that makes sandpaper gritty). Planets can also be shrouded by photochemical hazes arising from the effect of ultraviolet on methane and other hydrocarbon gases; these have much in common with the smoggy haze that plagues polluted cities as well as the orange haze of Titan's upper atmosphere.

Clouds exert a profound effect on a planet's climate. Many kinds of clouds, including water, ammonia, and silica (SiO_2) clouds, strongly reflect instellation back to space. Some kinds of clouds, however, have soot-like absorption and lead to intense local heating of the atmosphere. These include clouds made of SiO and metallic iron (think of your cast-iron frying pan, not a shiny polished stainless-steel saucepan). Thick clouds invariably have a strong effect on the infrared radiating level as well, since like most condensed substances they emit nearly like blackbodies. If they occur in a cold region of the atmosphere, they exert a potent greenhouse effect, which can offset or overwhelm the cooling effect of their reflection of instellation.

The presence of clouds can be detected in both transit depth spectroscopy and secondary eclipse emission spectroscopy, because the absorption and emission of clouds (like that of most condensed substances) depends much less strongly on wavelength than it does for gases. The presence of high clouds thus flattens the spectrum. In some cases, the spectrum is so flat that it can be difficult to distinguish a bare-rock planet from one with a cloudy atmosphere. This would be the case for the Neptini GJ1214b, were it not for the fact that its measured density is so low that it must have an extensive gaseous atmosphere. A great diversity of cloudiness, arising from a variety of condensed substances, has been observed amongst the exoplanets.

The habitable zone

Just knowing that there are billions and billions of planets out there increases the prospect that we are not alone in the Universe, but which of these planets could host life? The state of theoretical biology is such that it is hard to say anything definite about life as we don't know it, so attention naturally focuses on conditions that approximate those found on Earth. A common filter for habitability is the requirement that liquid water be able to exist at the planet's surface. That presumes, among other things, that the planet actually has a surface, although it could be an ocean surface without any rocky continents rising above the water.

This is a rather parochial notion of habitability, as it leaves out such possibilities as life in subsurface ice-covered oceans, life living deep in a rocky crust (which actually happens on Earth), or free-floating life on gassy planets without a distinct surface. Jupiter's moon Europa has an icy crust with a surface temperature that never rises above 100K, and essentially no atmosphere, but it still has a liquid subsurface water ocean that could potentially host life. Jupiter itself has a 'habitable zone' around the 18 atmosphere pressure level, where temperatures are Earthlike and liquid water can exist in the form of cloud droplets. In Larry Niven's novel, *The Integral Trees* a form of free-floating ecosystem was imagined that could live in such an environment. Venus, despite its torrid surface conditions, has a habitable zone within its cloud deck, which also provides a source of liquid water within the cloud droplets. The clouds, however, consist of concentrated sulphuric acid with acidity far in excess of the most acid environments to which life has adapted on Earth, so Venus cloud life would still have formidable barriers to overcome. Yet more speculatively, silicon-based life could be imagined surviving at high temperatures (think of the Horta in the famous Star Trek episode). Life could perhaps exist in chilly CO_2 or methane oceans. Not much can be said yet about the nature of such life and how to detect it, so here we'll stick to the conventional definition of the liquid-water habitable zone. Even so, when talking about the habitable zone, it is crucial to remember that actual habitability is contingent on a planet being able to maintain a suitable atmosphere. Our Moon is firmly in the Sun's habitable zone, but (without technological intervention) is not actually habitable. The Moon fails habitability because it is airless, but retaining too much atmosphere can also be a problem. Water can be liquid at temperatures up to 647K, but a liquid ocean at a temperature of 500K would not be habitable for life as we know it; a planet that retained a sufficiently thick hydrogen atmosphere could easily become that hot, and thus be uninhabitable even if it were equipped with a liquid water ocean.

Planets currently in the habitable zone of main sequence M stars were blasted with extremely high installation and atmosphere-eroding extreme ultraviolet radiation during the long pre-main sequence stage, and M stars put out a higher proportion of extreme ultraviolet even once they settle onto the main sequence. The problem of keeping the right sort of atmosphere may prove fatal to the habitability of M star planets, which may wind up as bare rockballs unless they start off with so much hydrogen that their atmospheres are uninhabitably thick once their stars settle down. We know of bare-rock M star planets like LHS3844b, and M star planets that have massive hydrogen envelopes like K2-18b, but so far none that have a 'just right' atmosphere. Observations of the coming decade will shed much light on this crucial question.

If a planet is too close to its star, and has too much installation it will be, in the words of Goldilocks, 'too hot'. If it is too far from its star, and has too little installation, it will be 'too cold'. Next, we'll take a look at just where in the range of installation a planet is 'just right' to keep Goldilocks' porridge liquid at the surface.

Let's consider the inner edge of the habitable zone. The energy balance controlling the climate here is sketched in Figure 18.

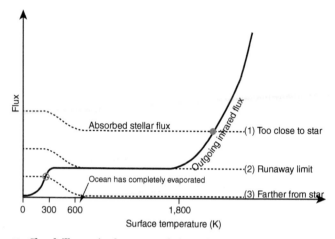

18. Sketch illustrating how energy balance determines the inner edge of the habitable zone. The solid line shows the rate at which the planet cools to space by emitting radiation while the dashed line shows the rate at which energy received from the planet's star is absorbed by the atmosphere and surface; equilibrium occurs where the two lines cross. Three different curves of absorbed stellar radiation are shown, corresponding to three orbital positions and their corresponding instellation values. The energy fluxes are plotted as a function of surface temperature; the absorbed stellar flux goes down with temperature because the higher atmospheric moisture content reflects more energy to space. For orbit (3) there is an equilibrium at a habitable temperature, but for position (1) the system runs away to over 1,800K.

For a planet with a fixed mass and composition of atmosphere, increasing its surface temperature increases the radiation of energy to space in accordance with the fourth-power law, so that the planet can always warm up enough to radiate away all the energy it receives from its star. If the planet has a surface ocean, however, increasing temperature causes more of the ocean to evaporate into the atmosphere, and since water vapour is a potent greenhouse gas, this feedback causes an additional boost to the warming. Once there is enough water vapour in the atmosphere, though, the temperature of the radiating level becomes controlled by the temperature at which water vapour at that level condenses, at which point the radiating temperature becomes independent of surface temperature. As a result, there is a limit to the rate at which a planet with a water ocean can radiate energy to space. If the rate at which the planet absorbs energy from its star—the instellation minus the proportion reflected back to space—exceeds this limit, then the surface warms inexorably until the entire ocean has evaporated into the atmosphere. For an ocean the mass of Earth's ocean, the surface temperature at that point is a lethal 600K, but unless the ocean is considerably less massive than our own, the planet must continue heating until the upper atmosphere becomes hot enough to begin radiating additional energy to space. At that point, the surface generally is hot enough to melt rock. The planet has lost a water ocean, but gained a magma ocean.

The steam-dominated atmosphere does not stick around forever. Exposed to intense ultraviolet light in the upper atmosphere, it breaks up into hydrogen and oxygen. The light hydrogen escapes to space, leaving oxygen behind, The oxygen may accumulate as an abiotic oxygen atmosphere. Earth's oxygen, in contrast, is produced by photosynthesis, which converts carbon dioxide, water, and sunlight to oxygen and biomass. Or, the abiotic oxygen may react with rocks, forming oxidized minerals such as Fe_2O_3, a form of rust. Some oxygen may also be dragged away to space by the escaping hydrogen.

The chief difficulty in determining the instellation threshold which defines the inner edge of the habitable zone is determining the proportion of the instellation that is reflected back to space—the albedo. Reasonable estimates can be made of the surface albedo of rocky or ocean-covered planets. Atmospheric gases also reflect some instellation, via the process called Rayleigh scattering which also accounts for the blue colour of Earth's sky. This, too, can be estimated, given atmospheric mass and composition. Clouds pose a much bigger challenge; water clouds reflect a great deal of instellation, but they can also have a substantial greenhouse effect which provides compensating warming. Even allowing for a fairly generous estimate of net cloud effects, the instellation threshold for the inner edge of the habitable zone for a Sunlike star is a mere 6% greater than Earth's current instellation. We are living perilously close to the edge. The redder light of stars cooler than the Sun, e.g. M stars, is not reflected as strongly by atmospheric gases, so the instellation threshold for such stars is somewhat lower than for Sunlike G stars. The greatest uncertainty in the maximum instellation compatible with habitability is nonetheless due to cloud behaviour, and this is likely to depend quite a lot on planetary characteristics such as rotation rate and atmospheric composition.

A viable theory for the hot, dry state of Venus is that it succumbed to a runaway greenhouse and lost its water sometime in the past several billion years. Venus today, with 1.9 times Earth's instellation, is well within the instellation zone where a runaway greenhouse would be expected, barring a strongly stabilizing cloud effect (which can't at present be absolutely ruled out). However, main sequence stars gradually get brighter as they age, as can be seen from Figure 8. Four billion years ago the Sun was about 30% less luminous, and Venus could possibly have supported a liquid water ocean. Given uncertainties in cloud effects, it is entirely possible that Venus remained habitable until relatively recently, if it started out with a water ocean. However Venus might also have lost its water in its youth, before it ever had a chance to condense into an ocean. Or, it may never have received significant amounts of water at all. Unravelling the history of Venus is one of the big tasks of planetary science, and it engages all the key questions surrounding the nature of the inner edge of the habitable zone. As we will see shortly, the dryness of Venus, however it happened, accounts for its thick CO_2 atmosphere. When it becomes possible to characterize the atmospheres of rocky exoplanets with Venus-like instellation, much will be learned about the variety of ways habitability may fail—or be preserved—for such planets. Good targets for such investigations include Trappist 1b, c, and d, which are known to be rocky, and have instellations 4.25, 2.27, and 1.14 times Earth's. The rocky planet GJ1132b, which receives 18.6 times Earth's instellation, will provide a counterpoint on the very hot side. What kind of atmosphere, if any, would such a planet have? The coming decade will provide answers to such questions.

Now we'll turn attention to the outer edge of the habitable zone. All other things being equal, the farther a planet is from its star, and the lower its instellation, the colder it will be. At some distance, it will become so cold that any water at the surface freezes, and any ocean will become completely ice-covered like the ocean of Jupiter's moon Europa. But why can't we just make up for low instellation by giving the planet an atmosphere with more greenhouse warming? To some extent we (or nature, at least) can, although there are limits to how much a planet can be warmed by the greenhouse effect, which depend on the greenhouse gas in question and the planet's instellation. Without the possibility of the greenhouse effect compensating for low instellation, the habitable zone would be exceedingly narrow, and a planet receiving just a bit less instellation than Earth would be frozen over.

The greenhouse effect that has maintained Earth's habitability for billions of years is a collaboration between CO_2 and water vapour. We have other greenhouse gases in our atmosphere, but we would do fine without them. CO_2 is a long-lived greenhouse gas that, for Earthlike temperatures, is not subject to rapid removal by condensation and rainout. Given a water ocean, the water vapour content of the atmosphere is largely determined by its temperature, since if too much water enters the atmosphere, it rains out. For a planet that is not in the runaway greenhouse zone, the temperature settles down to a value that is greater than it would have been for CO_2 alone, but well short of what would be required for the entire ocean to evaporate into the atmosphere. In defining habitability, it is natural to think first of the CO_2–water system that has nurtured life on our own planet.

There are two factors that can limit the maximum greenhouse effect. The more gas one puts into an atmosphere, the more instellation it reflects back to space. For Earthlike instellation, it takes only a small amount of CO_2, equivalent to about .03% of Earth's atmospheric mass, to keep the oceans from freezing over. For considerably lower instellation, the required amount of CO_2 would be the equivalent of ten times or more of the entire mass of Earth's atmosphere, and that would reflect back a great deal of the instellation. At some point, the reflection overwhelms the greenhouse warming and adding more CO_2 cools the planet instead of warming it. If the instellation is too low, the maximum greenhouse limit sets in before surface temperatures have become high enough to keep the ocean from freezing.

When instellation is high enough, when you stuff more CO_2 into an atmosphere it builds a thicker atmosphere. This is how Venus can have a nearly pure CO_2 atmosphere about a hundred times as massive as Earth's atmosphere, and be very hot as a result. Earth is also in a zone where CO_2 would accumulate were it not for slow geochemical processes that, in the presence of water, can remove CO_2 from the atmosphere. However, if the instellation falls below a certain threshold, as more CO_2 is added to the atmosphere it condenses out and accumulates on the surface as a CO_2 ocean, rather than adding to the greenhouse effect.

The joint effects of the two mechanisms limiting the CO_2 greenhouse effect are illustrated schematically in Figure 19. Taking both limits into account, a planet orbiting an M star requires at least 25% of Earth's instellation in order to remain unfrozen. A planet orbiting a Sunlike G star needs 35% of Earth's instellation and one orbiting a hot F star needs nearly 45%. Planets around cooler stars require less instellation to remain unfrozen because their stars put out more of their light as longwave infrared light, which scatters less efficiently from atmospheres than shorter wavelengths. The precise limits depend somewhat on the mass of the planet (or more precisely, its surface gravity). The stated limits

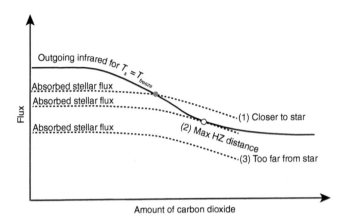

19. As for Figure 18, but for the outer edge of the habitable zone. In this case the surface temperature is held fixed at the freezing point, and the fluxes are plotted as a function of the amount of CO_2 in the planet's fluid envelope. The absorbed stellar flux goes down with CO_2 because increasing CO_2 increases reflection. For orbit (1) adding enough CO_2 can yield an equilibrium at the freezing point; to the right of the equilibrium, the planet is warmer than freezing. For orbit (3) no amount of CO_2 can warm the planet to the point where water ice melts.

do not take into account cloud effects, which could shift the habitable zone by an uncertain amount.

As one goes towards the outer edge of the habitable zone, progressively more CO_2 is needed to keep the planet unfrozen. At the outer edge, the amount required would bring the surface pressure up to about ten times Earth's surface pressure. In contrast, the amount of CO_2 in Earth's atmosphere accounts for only .04% of Earth's surface pressure. It's not much, but if it were removed, Earth's oceans would freeze over right to the Equator.

The deep carbon cycle and habitability

If the amount of CO_2 in an atmosphere were a matter of chance, the habitable zone would be severely constricted both in time and space. Few planets would have just the right amount to support surface liquid water. A planet near the inner edge, like Earth, could wind up with as much CO_2 as Venus, rendering it uninhabitably hot, but a planet near the outer edge might not get enough to keep it unfrozen. A planet around a type K or a hotter star that had the right amount of CO_2 in its youth, when its star was fairly dim, would find itself eventually too hot as the star aged along the main sequence—or a planet like Earth, that has enough CO_2 to support liquid water now, would have fallen into a globally glaciated Snowball state in its youth, from which it would have taken billions of years to recover. Fortunately, for a rocky planet made mainly of silicates, the amount of CO_2 in the atmosphere is determined by a balance between what comes out of volcanoes and what is taken out of the atmosphere by reactions between CO_2 and silicates, which form carbonate rocks such as limestone. This reaction only takes place in the presence of liquid water, and the removal rate increases with both temperature and rainfall rate (which also tends to increase with temperature). This magical CO_2–water–silicate–carbonate system acts as a thermostat, which operating over millions of years tends to adjust the surface temperature to favour the presence of liquid water at equable temperatures.

In order for the silicate weathering thermostat to maintain habitability, however, there must be a means of recycling the carbonate formed in the crust by engulfing it into the deep interior of the planet where the temperature is high enough to decompose it and release CO_2 back into the atmosphere in the form of volcanic outgassing. Without a recycling process, all the CO_2 would eventually get bound up in crustal carbonates and the planet would enter a deep freeze. This is where long-lived radioactive elements, notably uranium-238, thorium-232, and potassium-40, enter into the story of habitability even though they make up only a tiny fraction of a rocky planet's composition. The half-life of thorium-232 is fourteen billion years, and that of uranium-238 is 4.5 billion years. Potassium-40 has a shorter half life of 1.2 billion years, but because it is relatively abundant it is still providing a significant amount of the heat that keeps Earth's volcanism going, even 4.5 billion years after the formation of our planet.

The heat produced by the decay of these elements sustains the dynamic processes that feed volcanism and crustal recycling. On Earth, this is manifest in the form of plate tectonics, where carbonates that are washed into the ocean and deposited on the ocean floor are subducted as great slabs, to be replaced by new ocean floor created at the mid-ocean ridge. Some of the CO_2 is released locally near subduction zones through arc-volcanism, but the part that is subducted into the deep interior outgasses at the mid-ocean ridge and elsewhere. The conditions under which plate tectonics occurs have not yet been resolved. Is plate tectonics favoured by a water ocean? Are rocky planets larger than Earth more amenable to plate tectonics or less? There are abundant theories, but not yet any definitive answers. It is also possible to recycle crustal carbonates without plate tectonics, so long as there is sufficient volcanism. Progressively flooding the surface with lava pushes older crust down into the interior, where it can be hot enough to melt and release CO_2.

The amount of atmospheric CO_2 permitted by the balance of sources and sinks can impose a much more severe limit on the outer edge of the habitable zone than the maximum greenhouse or condensation limit. Outgassing, crustal recycling, and silicate weathering are complex processes, very imperfectly understood even for Earth, and things may play out very differently amongst the wide range of compositions and surface conditions encountered on rocky exoplanets. When atmospheres of rocky exoplanets come into view, an examination of the way atmospheric CO_2 content varies with instellation will help determine how prevalent the Earthlike way of maintaining habitability is.

It is rather remarkable that the whole apparatus of nucleosynthesis, generation of long-lived radioactive elements, and the chemical constants that determine the freezing point of water and the properties of the silicate weathering reactions have conspired to permit the operation of the silicate weathering thermostat. The 'anthropic' principle would state that of all possible Universes, things have worked out this way because a Universe has to have something near these characteristics in order to allow us to be here to notice such things A less anthropic—and probably more humble—view is that we evolved to take advantage of this particular characteristic of our Universe, and that other forms of life could evolve to make use of other geochemically stabilized habitats.

The thick CO_2 atmosphere of Venus which renders it so hot did not arise because the planet was endowed with more carbon than Earth. Rather, in the absence of liquid water at the surface—either because Venus lost its water through a runaway greenhouse or because it was born dry—there is no way to remove the CO_2 produced by volcanic outgassing, so it all winds up in the atmosphere. Something similar would happen to Earth if it lost its water.

The science of habitability is still in its infancy, and we rely too much on knowledge of the way things work on Earth. Other gases, including hydrogen, could extend the outer edge of the habitable zone far beyond what is permitted by the CO_2-water supported greenhouse. Mechanisms analogous to the silicate weathering thermostat are not known for these gases, but that may just mean that we haven't been thinking about such things long enough, or been clever enough. Time will tell.

When it comes to habitability, small—but not *too* small, is beautiful. Planets that are too big are likely to have a massive gaseous envelope which renders the surface (if any) inhospitable, but can a planet be too small to be habitable? If a planet is too small, its weaker gravity makes it harder to hold on to an atmosphere. Certainly, the 2km asteroid that was the home of

Le Petit Prince of Antoine de St. Exupéry's book could not have kept an atmosphere, but even a planet the size of Mars can lose most of its atmosphere, and evidently did. Heat escapes from small planets more easily than from large planets, because small planets have a greater ratio of surface area (through which heat escapes) to volume (which stores heat and generates it by radioactive decay). A small planet would require a much greater amount of radioactive heat generation, per kilogram of rock, to maintain vigorous lava-producing volcanism. This is basically for the same reason that a mouse needs to consume many more calories per gram of body weight than a human. A typical mouse, weighing 40 grams, consumes 10 kilocalories (sometimes called simply 'calories' in everyday usage) per day. Scaled up to the mass of a human, that would amount to 25,000 kilocalories a day—equivalent to 7kg of dry spaghetti.

Volcanism can nonetheless crop up in surprising places. Jupiter's moon Io, although only 3,600km across, is one of the most volcanically active bodies in the Solar System, owing to the huge amount of tidal heating deposited by the flexing of the body caused by Jupiter's gravitational field. Volcanism based on melting ice requires little energy, and can be sustained on a quite small body. So, tiny Ceres, at only 1,000km across the largest asteroid, is able to muster enough internal heat to create outbursts of muddy cryovolcanism.

Mars provides an important window into what happens to a planet near the outer edge of the habitable zone, and it poses many puzzles that are still unresolved. With an instellation of .43 Earth units, it is well within the habitable zone today. Its habitability has failed because it somehow lost most of its atmosphere. The giant volcano Olympus Mons may have been active as recently as 115 million to two million years ago, but evidently there has not been enough volcanic outgassing to regenerate a thick CO_2 atmosphere. However, abundant geological features on Mars, some as old as four billion years, indicate that in its youth Mars, at least episodically, supported great rivers of running water at its surface. At that time, when the Sun was 30% less luminous than today, Mars was well beyond the outer edge of the CO_2/H_2O-supported habitable zone. There must have been some additional source of greenhouse warming in the atmosphere at the time, but the book is still open on just what that was.

How many habitable planets are there?

From Figure 12 it is evident that there are a handful of planets with instellation in the habitable zone range, and which are also of a size or mass small enough to have a potentially rocky composition. Most of these orbit M stars, but that is just because smallish planets with Earthlike instellation are easier to detect around low mass stars. Taking into account the effect of stellar type on the habitable zone instellation boundaries, at the time of writing there are eighteen known planets in the habitable zone, including Proxima Centauri b (orbiting our nearest stellar neighbour) and planets d, e, f, and g in the remarkable Trappist 1 system. There are an additional twenty-six near-misses which could perhaps be rendered habitable if cloud conditions or other uncertain bits of climate physics become more favourable than current best estimates. One would like to extrapolate from this number to an estimate of the proportion of all stars that have a planet in their habitable zone.

This quantity, called eta-Earth (ηE) is the holy grail of planetary population statistics. It is difficult to estimate from the handful of habitable zone planets we know, because of the need to correct for observational bias—i.e. that potentially habitable planets are much harder to detect than thoroughly uninhabitable ones like Hot Jupiters. Most estimates also involve extrapolation from occurrence rates of observed (but uninhabitable) planets to rates for habitable zone (but so far unobserved) planets. The estimated value of ηE also depends on whether it includes M stars, or just more Sunlike G or K stars, as well as where one draws the line on the habitable zone. Regardless of the fine points of the estimate, the evidence so far is that habitable zone planets are *not rare*. For M stars, ηE estimates range from 25% to nearly 50%, and for G and K stars the best current estimates are between 37% and 88%. Taking one of the lower estimates of the number of stars in the Milky Way, that's at least 185 million habitable zone planets around G and K stars, and twenty-two billion around M stars in our galaxy alone. That's a lot of potentially habitable real estate.

But just because a planet is in the habitable zone doesn't mean it's actually habitable. It still needs a suitable atmosphere. For the most part, it is not even known if the habitable zone planets discovered so far have any atmosphere at all; in the few cases where an atmosphere has been detected the indications point to a thick hydrogen atmosphere that would render the surface uninhabitably hot. There is not yet any basis for estimating what proportion of habitable zone planets have suitable atmospheres. The veil on this important question will lift in the coming few decades. Because M stars are so numerous, if it turns out that M star planets can commonly have suitable atmospheres, the Universe is surely teeming with life.

Chapter 7
How it all ends

In the Bhagavad Gita (11:32), Lord Krishna says *Time I am, destroyer of the worlds*. Time provides us with a river of worlds, but also brings each of them to an end, each in its time. Here we'll look into the ways this end can come about.

Fire

Some six billion years from now, if anybody is left standing on Earth to witness it, they will see the Sun begin to go through some alarming transformations. The beginning of the end comes when hydrogen fuel is exhausted in the core. The core is not yet hot enough to ignite fusion of helium into carbon, so fusion in the core ceases. At first this will be noticed only by Solar physicists, who will see a drop in the flux of neutrinos—nearly undetectable particles that are a by-product of fusion, which interact so weakly with matter that they can stream through the outer envelope of the Sun virtually unimpeded. No longer supported by fusion heat, the star begins to contract, heating up the interior sufficiently for unexpended hydrogen to begin fusing in a shell surrounding the core. The Sun has entered the subgiant stage. In this stage, luminosity increases much more rapidly than it does during the main sequence stage. After a half billion years it will have increased so much that the inner edge of the habitable zone will have moved out beyond Mars to the inner Asteroid Belt, and the outer edge extends nearly to the orbit of Jupiter. Our own Solar System has no bodies in the band primed to become habitable, but other planetary systems may see a brief window of habitability open up for formerly frozen rocky planets.

About a hundred million years later, things really get bad. Still on the subgiant branch, the Sun inflates to reach nearly Earth's orbit. Its surface gets cooler and glows a dull red, but the vast increase in surface area makes up for that and the luminosity increases to nearly a thousand times that of its value at the end of the main sequence. Mercury and Venus have been engulfed, and even if Earth escapes this fate its instellation increases to the point where it is stripped of volatiles. The rocky surface melts into a magma ocean, which produces a rock vapour atmosphere. Some of that escapes to space, but if the planet avoids being engulfed, it probably won't entirely evaporate, although it will be thoroughly cooked. Can things get any worse? Indeed they can, for about this time the core of the star gets hot enough to ignite fusion of helium into carbon, which happens in a flash during which the luminosity increases by another one or two orders of magnitude. The star has entered its red giant stage. For stars near Solar mass, there is usually a second helium flash around a hundred million years after the first, whereafter the outer envelope of the star, consisting of a third or more of its mass, is blown off. Not long afterwards, the star evolves into a hot but small, low luminosity white dwarf hardly bigger than the Earth, which gradually cools and fades away. The luminosity is less than a hundredth of the peak main sequence luminosity, and any planets formerly in the habitable zone are locked in a deep freeze—if they have any volatiles left to freeze.

The same scenario plays out, with some variations, for planetary systems around all stars with a mass from around 0.5 to 8 times the mass of the Sun. More massive stars end their lives in a searing supernova explosion (passing first through a red giant stage if their mass is not too great), although such stars are so short-lived that their planets are unlikely to be inhabited. We nonetheless have them to thank for helping to seed the Universe with the uranium, thorium, and potassium-40 that provide the heat needed to sustain outgassing and keep rocky planets habitable. Even if current research indicating that most uranium and thorium come from neutron star mergers or collapsars proves correct, such objects are the remnants of supernova explosions of massive stars, and so they rely on a good rate of production of short-lived stars.

The time of leaving the main sequence and the duration of the subgiant and red giant stage depend strongly on mass, and also somewhat on the chemical composition of the star. A 1.4 Solar mass F star will live for under five billion years on the main sequence before turning into a red giant; if our planet were orbiting an F star, we'd be gone by now. A .6 Solar mass K star will sustain habitability for nearly a hundred billion years—nearly seven times the current age of the Universe. Towards the higher end of the mass range, helium fusion ignites more smoothly, without a helium flash. Towards the lower end of the mass range, helium fusion never ignites at all, but the hydrogen shell fusion nonetheless results in an inflated high luminosity reddish star that closely resembles a helium-powered red giant. Regardless, at the end of the main sequence for these stars, planets formerly in the habitable zone are fried in an inescapable habitability crisis, while farther out planets enjoy at best a cosmically brief summer of habitability.

Many of the brightest stars in the night sky are subgiants or red giants. Arcturus, with a mass just 8% greater than that of the Sun, is currently ascending its subgiant branch and has puffed up to twenty-five times the Sun's radius and reached 170 times its luminosity. It provides a nightly reminder of our ultimate fate. We may take comfort though that such stars puff out life-giving elements like a dandelion spreading its seeds, paving the way for the reincarnation of habitable worlds.

Actually, there is an important part of the story of the end times we have swept under the rug: few planets that start in the habitable zone will be able to live out the full main-sequence lifetime of their stars, at least for stars in the mass class discussed above. Recall that stars get gradually brighter throughout the main sequence (see Chapter 4). Earth is currently quite near the inner edge of the habitable zone, and will cross the runaway greenhouse threshold in about a half billion years, whereafter the oceans will evaporate away, the surface will be heated to sterilizing temperatures (possibly even creating a transient magma ocean), whereafter water will break up and be irretrievably lost to space. Then, we will turn into Venus, with a thick CO_2 atmosphere, as silicate weathering ceases. This is a habitability crisis all planetary systems around sufficiently massive main sequence stars confront, though planets that begin life near the outer edge of the habitable zone will be able to live out a somewhat greater proportion of their host star's main sequence lifetime. This represents a terrible waste of a lot of perfectly good main sequence lifetime. However, unlike the subgiant crisis, which can't be addressed short of pulling up stakes and moving elsewhere, there are technological fixes that would be readily accessible to any advanced civilizations that may exist at that time. All that is needed is to reflect a bit more instellation back to space, or divert it so that it scoots past the planet. Diverting 10% of the stellar flux would be enough to stave off the runaway greenhouse for about a half billion years. Using technology that is almost within reach today, this could be done by injecting reflecting particles into the upper atmosphere, although such a climate hack has the disadvantage that it would need to be renewed annually, to replace particles that fell out of the atmosphere. There are those who would like to deploy such albedo hacking right now as a response to the global warming crisis, rather than facing up to the challenge of decarbonizing the world economy. Such an action would be foolish in the extreme not least because the carbon dioxide that human activities inject into the atmosphere has warming effects that extend over millennia, whereas stratospheric aerosols fall out of the atmosphere in a year or two. Because of its long atmospheric persistence, carbon dioxide will continue to accumulate in the atmosphere so long as industrial emissions continue, requiring ever-greater amounts of albedo hacking, and if there is ever a termination of this form of planetary life support, it would result in a global catastrophe in which centuries of pent-up warming were unleashed in a matter of a decade. It would put our planetary home in a precarious state, like living perpetually under the Sword of Damocles. This would be tragic, given that achieving the eminently achievable goal of a low-carbon economy would leave us in a much more resilient future.

Albedo hacking is a bad response to human-caused global warming, but if the alternative is a runaway greenhouse induced by stellar brightening, the calculus of risk is different, as there are no other options to stave off the catastrophe. A solution available to a somewhat more advanced civilization than ours would be to build a diverging lens in space, which would spread out incident radiation so that a sufficient portion was diverted from absorption by the planet (a diverging lens is better than a mirror since it requires less mass and creates less light pressure that would tend to push it off-station). By increasing the diversion up to half of the incident flux as the star brightens, the runaway greenhouse can be averted out to the end of the main sequence stage.

The Solar System doesn't have planets waiting in the wings to become habitable when Earth passes the runaway greenhouse threshold. Mars is too small and has lost its atmosphere and ability to regenerate it by outgassing. Planets never formed in the asteroid belt, and none of the Jovian moons would provide a suitable habitat. Other planetary systems, however, probably have Snowball planets with volatiles locked in ice and enough interior heating to maintain a habitable climate after their stars brighten enough to bring on a planetary springtime. Such planets around G stars offer the possibility of 'second half' habitability, and for K stars, which brighten slowly over a hundred billion years, there can be successive outward expansions of the habitable zone, giving new meaning to Teilhard de Chardin's concept of ever-expanding waves of consciousness.

Ice

Planets around low mass M stars will not suffer a crisis by fire, either from runaway greenhouse or a red giant catastrophe, because such stars evolve very slowly. Our nearest neighbour, the red dwarf Proxima Centauri (which has a planet in its habitable zone), will still be shining with more or less its current luminosity a trillion years from now. Nonetheless, such planets do not get to remain habitable forever. Because thorium-232, the longest lived radioisotope maintaining the ability to recycle volatiles that get bound up in the crust, has a half life of just fourteen billion years, it is hard to keep CO_2 recycling going for more than a hundred billion years, or even less if the planet doesn't have plate tectonics. When outgassing winds down, unchecked silicate weathering draws CO_2 out of the atmosphere. Such planets end their lives in ice, not fire.

However, if a technological civilization emerges and survives on such a planet before it freezes out there is an easy fix to maintain habitability more or less indefinitely, because it takes very little energy to cook CO_2 back out of limestone at a rate sufficient to offset silicate weathering. In fact, our civilization does that already at sufficient scale, in the process of cement production—and that just counts the CO_2 directly released from the limestone, not the CO_2 released by fossil fuels burned to provide the energy. So, something as simple as Solar-powered cement production at a scale our economy has already achieved could keep an M star planet habitable for a trillion years.

Dark matter is what holds our galaxy together, but it can't make stars and planets; that is up to the 10% or less of our galaxy that is baryonic matter. Of this fraction about a third is bound up in stars.

The ratios probably differ from galaxy to galaxy, but the general picture is that baryonic matter is a minor constituent, and that a substantial fraction resides in stars at any given time. Matter that finds itself in massive stars is quickly recycled into the pool available for forming new stars, and even a G star like the Sun only holds onto its mass for ten billion years or so. As time goes by, hydrogen and helium are irreversibly converted to heavier elements so that the metallicity of the Universe increases; perhaps it will eventually become possible to form rocky giant planets. An even bigger factor in galactic evolution, though, is that once matter finds its way into an M star (or a white dwarf or black hole) it is stuck there basically forever, eventually starving the galaxy of the hydrogen needed to form new stars. Given that most stars are M stars, star formation could cease in as little as a trillion years, though some estimates put the date considerably farther off. But most of the M stars made during that time will still be shining for a long time to come. Their formerly habitable worlds will have long frozen over because of depletion of the radioactive heating that maintains recycling of greenhouse gases. The Universe will have become a dilute soup of dim galaxies, each isolated in its own bubble of darkness. If civilizations arise in time and persist, some of these M star worlds will be able to maintain habitability through technological intervention, assuming such civilizations still feel the need for planets.

Worlds enough, and time

It is *triste* to contemplate the winding down of the Universe into a cold, dark, lonely place, but we are a young species in a young Universe, with vast reaches of time before us. It is certainly true that there are countless worlds out there that could support life as we know it, and probably countless more that could support life as we don't know it. It may be that the Universe is teeming with life waiting to make our acquaintance. Or, we may well be the *first ones* in our galaxy to make the leap to sentience. The vast distance between stars poses a severe barrier to individuals or even societies making the journey. Protoplasm is just too fragile and short-lived a medium to be up to the task of such voyaging. However, at a tenth the speed of light, the whole galaxy can be traversed in a million years. That's a long time for protoplasm, but it is not a stretch to think of the data that makes us what we are—embodied perhaps in silicon or some other sturdy information-bearing material and reconstituted at destination—spreading throughout the galaxy, hopping from planet to planet along the way like Pacific Islanders in their canoes. If life—or complex life—is rare, it may well be our destiny to seed the Universe with an expanding wave of consciousness. But it is to be hoped that we will leave abundant worlds alone to develop their own destinies. There are worlds enough, and time.

Regardless of our destiny, the clear miracle is that little blobs of protoplasm making up a species barely a hundred thousand years old living in the outskirts of a not especially remarkable galaxy have been able to learn so much about the Universe around us. We have peered back to the moments after the Big Bang, and have inferred the likely fate awaiting trillions of years from now. We have been able to probe the farthest reaches of the Universe by detecting the feeble vibrations of gravitational radiation, and have begun to lift the veil on what planets are out there, and what they may be like. The saga of exploring planetary systems has just begun. There is no limit to what we can accomplish, if we can make it through the next few hundred years without crashing the Earth's habitability, and without letting the authoritarianism emerging throughout the world crush the human spirit, dividing us one from the other, and separating us from our better natures.

References and further reading

There are a number of books in the *Very Short Introduction* series which further illuminate a number of the topics touched on in the present work. These include the *Very Short Introductions* to Stars, Black Holes, Galaxies and Astrobiology. The following list includes some of the sources from which the material in the present book was drawn together with some suggestions for further reading.

John Bahcall (2021). How the Sun shines. NobelPrize.org. Nobel Media AB. https://www.nobelprize.org/prizes/uncategorized/how-the-sun-shines-4. This Nobel Prize lecture gives an outstanding overview of the processes that make stars shine, and the history of the discovery of these processes.

Gregory L. Baker (1983). Emanuel Swedenborg: An 18th century cosmologist. *The Physics Teacher* 441–6.

A.C. Bhaktivedanta Swami Prabhupada (2006). *Bhagavad Gita As It Is*. Thatcham: Intermex Publishing Ltd., 810pp. This is the translation from which the quote at the beginning of Chapter 7 is drawn. The quote is best known in English from the form J. Robert Oppenheimer used upon witnessing the first atomic bomb detonation, which begins 'Now I am become Death . . . ' but the original Sanskrit word is more accurately translated as 'Time' according to most scholars.

Clare Dobbs (2013). Giant Molecular Clouds. *Astronomy and Geophysics* 54: 5.2–8.

Richard B. Larson (2003). The physics of star formation. *Reports on Progress in Physics* 66: 1651–97.

James Kasting (2012). *How to Find a Habitable Planet*. Princeton, NJ: Princeton University Press, 360pp.

Raymond T. Pierrehumbert (2010). *Principles of Planetary Climate*. Cambridge: Cambridge University Press. This book is for readers who want a deep dive into the subject of planetary climate. The early chapters are accessible to the general reader with the equivalent of A-level familiarity with physics and maths.

Tom Ray (2012). Losing spin: The angular momentum problem. *Astronomy and Geophysics* 54: 5.19–22.

Sean N. Raymond and Alessandro Morbidelli (2020). Planet formation: Key mechanisms and global models. *Lecture Notes of the 3rd Advanced School on Exoplanetary Science*, ed. L. Mancini, K. Biazzo, V. Bozza, and A. Sozetti. New York: Springer, 100pp. Available at: https://arxiv.org/abs/2002.05756

Karel Schrijver (2018). *One of Ten Billion Earths*. Oxford: Oxford University Press, 480pp.

D.M. Siegel, J. Barnes, and B.D. Metzger (2019). Collapsars as a major source of r-process elements. *Nature* 569(7755): 241–4. This article describes recent research on the role of collapsars and neutron star mergers in creating r-process elements such as uranium and thorium.

Arthur Stinner (2002). Calculating the ages of the Earth and Sun. *Physics Education* 37: 296–305.

N.A. Teanby, P.G.J. Irwin, J.I. Moses, and R. Helled (2020). Neptune and Uranus: Ice or rock giants? *Philosophical Transactions of the Royal Society A* 378(2187): 20190489. This article discusses the possibility that Uranus and Neptune may be rocky planets with an extensive hydrogen envelope rather than ice giants as conventionally pictured.

Pierre Teilhard de Chardin (1959). *The Phenomenon of Man*, trans. Bernard Wall. New York: Harper and Bros. Teilhard de Chardin was a Jesuit philosopher, theologian, and paleontologist. His work is not widely read today, but his thoughts on sentience and human destiny are well worth study. A number of recent reprints of his book are available, for example the 2008 Harper Perennial edition.

Jonathan P. Williams and Lucas A. Cieza (2011). Protoplanetary disks and their evolution. *Annual Reviews of Astronomy and Astrophysics* 49: 67–117. This article provides a somewhat technical but still fairly accessible overview of protoplanetary disks, both from a theoretical and observational perspective. It is pre-ALMA, but a post-ALMA review of the subject is not yet available.

Index

A

abiotic oxygen 111
absorption spectrum 82
aggregates 4–5
albedo 97, 102, 111
    hacking 125–6
ALMA (Atacama Large
    Millimeter/Submillimeter
    Array) 26–30, 38
ammonia (NH_3) 4
angular momentum problem 33–7
anthropic principle 118
Arcturus 124
Aristarchus of Samos 76
AS209 29
astronomical observations 22–6
    using ALMA 26–30, 38
atmospheres 66–75, 118
    clouds 106–7
    identification on exoplanets 82
    lost 119
    and temperature 100–5
atomic mass 57–8, 61–2

B

baryonic matter 1–2, 127–8
Bethe, Hans 9
Bhagavad Gita 122

Big Bang 1
binary neutron stars 53
blackbody radiation 17–21, 63, 97
black holes 52
Bondi radius 45–6
Bruno, Giordano 76–7

C

carbon 4
    outgassing 70
    on rocky planets 62–3
carbon dioxide (CO_2) 4
    critical temperature 73
    as a greenhouse gas 102, 105,
        114–15
    and habitability 114–18
carbon fusion 51
carbon monoxide (CO) 21
    depletion 29
    detection 30
centrifugal force 35, 37–8
Ceres, volcanos 119
Chajnantor Plateau 26
CI chondrites 62, 67–8
climate 107
clouds 106–7
collapsars 53
compression of gas 2, 3
Contact (1997 film) 11

core-collapse supernovae 52, 53
core segregation 64–6
Corot 7b 100
critical temperature 73
cross-sections 43

D

dark energy 1
dark matter 1–2, 127–8
Darwin, Charles 8
deep carbon cycle 115–20
density 86–8
digital imaging systems 22
Doppler shift 30, 79
DSHARP survey 28–9
dust grains 4–5

E

Eagle Nebula 11, 12
Earth
  age 9
  carbon vs. silicon 62–3
  clouds 106
  density 86–7
  end of 122–3
  escape velocity 68
  formation 9–10
  greenhouse effect 101–3
  instellation 114–15
  mass 85–6
  neon 71
  nitrogen 71
  outgassing 70–1
  temperature 96
  uninhabitable 125–6
earthquakes 37
eclipse depth 82
Eddington, Arthur 9
Einstein, Albert 9
electric fields 15–16
electromagnetic radiation 20–1, 97
electromagnetic waves 16–17
electromagnetism 15–17

elements
  heavier than iron 52–4
  mixing 54
  radioactive 54, 66, 116–17
  volatility 60–1
energy
  emission 19
  and temperature 97–9
  and wavelength 19
enstatite ($MgSiO_3$) 4, 61–2
enzymes 53–4
escape velocity 68–9
eta-Earth (ηE) 120–1
Europa 108
evaporation 59
event horizons 52
exoplanets 41, 62–3
  categorization 83–8
  core segregation 66
  early theories about 76–7
  habitability 120–1
  identification 78–83
  naming 77
  orbital periods (years) 88–90
Extremely Large Telescope 27

F

fayalite (Fe_2SiO_4) 65
51-Pegasi b 79–80
55-Cancri 77
55-Cancri e 100
fission 51
forsterite (Mg_2SiO_4) 65
free fall time 32–3, 45
freezing points 58–62
friction 36–7
fusion 8, 9, 51

G

Gaia mission 76
galaxies
  formation 2
  metallicity 58

Gamma Cephei A 79
gas
  accretion 64
  compression 2, 3
  formation 4–5
  gravitational accretion 44–6
  greenhouse effect 101–6
  in pebble accretion 43–4
  spectral fingerprints 82
gas giants 5, 44
  discovery 79–80, 84
  escape velocity 68
Giant Molecular Clouds 2–3,
    10–11, 32–3, 39
  angular momentum problem 34
GJ1132b 113
GJ1214b 87–8, 106, 107
globular clusters 10
Grand Tack model 94–5
graphite 4
gravitational accretion of gas 44–6
gravity
  in black holes 52
  and cross-section size 43
  in Giant Molecular Clouds 2–3,
    32–3
  and orbital frequencies 94
  in protoplanetary disks 37–40
greenhouse effect 101–6
  averting 125–6
  and habitability 113–15

H

habitability 108–21
  end of 125–8
  maintaining 127
HD143006 29
HD163296 28–9, 49
heat energy 2
helium 1, 9, 47
  in red giants 51
Helmholtz, Hermann 8
Herbig Ae stars 29
Herschel, Frederick William 15

Hertz (Hz) 16
Hill radius 45, 46
Holmes, Arthur 9
hot Jupiters 79–80, 84, 106
  clouds 106–7
Huxley, Thomas 31
hydrodynamic escape 69
hydrogen 1, 47–8
  accretion 5
  critical temperature 73
  dissociation 6–7
  formation of molecules 4
  fusion 8, 9
  in Giant Molecular Clouds 2
  in primordial atmospheres 67
  in protoplanetary disks 40
  surface of planets 73
  and ultraviolet 21
hydrogen degeneracy 88

I

ice 1
  density 62
  formation 4
  I, VI, and IX 72
  snowline 63–4
ice giants 5–6, 64
infrared 15, 16
  astronomical observations 22,
    23–6
  for identification of
    exoplanets 82–3
  and temperature 101–3
Infrared Astronomical Satellite
    (IRAS) space telescope 24
inner rocky planets 9–10, 44
  composition 62–3
instabilities 36
instellation 83–4, 97–9
  and habitability 109, 112–13
insulation 102
Integral Trees, The (Niven) 108
interferometry 27
Io, volcanos 119

iron 4, 51, 61–2
core segregation 64–6

J

Jupiter
angular momentum 35
clouds 106
density 88
escape velocity 68
formation 5
gas accretion 46
Grand Tack model 94–5
habitable zone 108
hydrogen surface 73
instellation 84
moons 92
snowline 63
temperature 105–6

K

K2-18b 109
K2-138 system 94
K2-141b 100
Kant, Immanuel 32
Kelvins 6, 18
Kepler-1000 77
Kepler space telescope 80
Kirchhoff, Gustav 20
Kuiper Belt 6

L

Laplace, Pierre-Simon 32, 34
Late Veneer 69–70
lava planets 100
LHS3844b 109
light
speed 15–16
theories 20–1
visible 14–15
liquid surfaces 72, 73
lithium depletion 25

luminosity 48–50, 54–6
Lyell, Charles 8

M

magma oceans 65
magnesium 4
magnetic fields 15–16
main sequence stars 47–50
mantle 65–6, 70
Mars
clouds 106
formation 9–10
habitability 119–20
mass
accretion 28–9, 35
calculating 80
categorization of planets 83–4, 85–6
and hydrodynamic escape 69
mass–energy equation ($E = mc^2$) 9
matter 127–8
emission of electromagnetic radiation 17
Maxwell, James Clerk 15, 20
Mercury
formation 9–10
orbital period 88
merger of stars 53
Mestel, Leon 36
metallicity 58, 128
methane (CH_4) 4
microwaves 16, 26–30
migration 94–5
Milky Way 1, 12, 54
habitable planets 121
minerals, formation 4
mini Neptunes 85
molecules, formation 4
monochromatic radiation 17
Moon
formation 10
habitability 109
temperature 97
tide-locked 90

moons 84–5
  of Jupiter 92
moving groups 10
music 91–2
mysticism 31–2

N

nebular hypothesis 31–2
neon fusion 51
Neptune
  clouds 106
  formation 5–6
  gas accretion 46
  ice 64
Neptune Desert 85
neutron stars 51–2, 53
New Horizons spacecraft 6
Newton, Isaac 32, 80
nitrogen 4, 51
  molecular (N_2) 71
Niven, Larry 108
noble gases 71
nucleosynthesis 51, 52, 57–8,
  117–18

O

oceans 72
  and habitability 110–11
olivines 61, 65
Oort Cloud 6
open clusters 10
Ophiuchus N3a star-forming
  region 29
orbital frequencies 92–5
orbital periods (years) 88–90
ordinary matter *see* baryonic matter
Orion Nebula 10–11
outgassing 66, 70–1
oxygen
  formation of molecules 4
  and habitability 111
  reaction with iron 66
oxygen fusion 51

P

parallax 76
particle theory 20
pebbles
  accretion 43–4
  formation 41
phase boundaries 73
phase curve 82–3
photo-evaporation 40, 68
photographic plates 22
photons 20–1
photosphere 18
Pillars of Creation 11, 12
Planck function 17–19, 21
planetary embryos 9–10
planetary systems 13
planetesimals 5, 9
  accretion of gas 44–6
  core segregation 65–6
  growth 42–4
  primordial atmosphere 67
planets
  angular momentum
    problem 33–7
  becoming uninhabitable 125–8
  calculation of mass 80
  density 86–8
  escape velocity 68
  formation 41–2
  mass 83–4, 85–6
  orbital frequencies 92–5
  radius 85, 87
  size 83–5, 118–19
  surface 71–5
  temperature 96–106; *see also*
    temperature
  tide-locked 90–1
plate tectonics 117
Pleiades (Seven Sisters) 11
Pluto 6
potassium 53, 66, 116–17, 124
power 19
primordial atmosphere 67
prisms 15

Index

protoplanetary disks 4, 22–5, 27
    snowline 63–4
    structures 37–40
    volatility of substances 59–60
protoplasm 129
protostars 3–4, 22–6, 28–9
    angular momentum 36
    snowline 63–4
proto-Sun 3, 6–8
Proxima 1b 89
Proxima Centauri 6, 89, 120, 127
PSR B1257+12 79
Pythagoras 91–4

Q

quantum theory 20–1

R

radial velocity (RV)
    technique 78–80
radiating temperature (T_{rad}) 102–3
radioactive elements 54, 66, 116–17
radio waves 15, 16
radius 85, 87
rapid neutron capture-process
    (r-process) 52–3
Rayleigh scattering 111
red dwarf stars 56
red giant stars 51–2, 123, 124
red light 18
refractory substances 59, 61
resonant chains 94–5
Ritter, Johann Wilhelm 15
rocky planets 9–10, 44
    atmospheres 66–75
    composition 62–3
    plate tectonics 117
    temperature 97–101
rotation rates 35
r-process (rapid neutron
    capture-process) 52–3

runaway greenhouse
    threshold 125–6
RV method 78–80

S

Sagan, Carl 81
sapphire in clouds 107
saturation vapour pressure 59
Saturn
    angular momentum 35
    clouds 106
    density 88
    formation 5
    gas accretion 46
    Grand Tack model 94–5
scale height 38
secondary atmosphere 69–70
Seven Sisters (Pleiades) 11
sharp phase boundaries 73
sidereal days 90
silicate (SiO_3) 4, 61
    dissolving with hydrogen 73–5
silicate mantle 65–6, 70
silicate weathering 117–18
silicon 4
    on Earth 62–3
silicon fusion 51
single pixel astronomy 81–2
Sirens of Titan (Vonnegut) 72
size of planets 83–5
    and habitability 118–19
slow process (s-process) 52, 53
snowline 63–4
Solar System 1
    composition 5–6
    as example of a planetary
        system 13
    formation 2–3
    Grand Tack model 94–5
    mass distribution 35
    mixture of elements 54
    nebular hypothesis 31–2

solid–gas phase transition 58–62
sound frequencies 91–2
spectral energy distribution 17,
    22–4
spiral arms 2, 58
Spitzer space telescope 24–5
s-process (slow process) 52, 53
stars
    classifications 49
    death of 122–5
    formation 11–12, 128
    luminosity 48–50
    main sequence 47–50
    naming 77–8
    neutron 51–2, 53
    red dwarf 56
    red giant 51–2
    white dwarf 50
Stefan–Boltzmann Law 19
stellar explosions 52
stellar remnants 50, 53
stellar wind erosion 68
stellar wobble 78–80
streaming instability 42–3
structures, formation 2
subgiants 50, 123, 124
sublimation 59
submillimetre waves 16–17, 26–30
Sub-Neptunes 88, 105
Sun
    age 8–9
    as a blackbody 20
    composition 56–8
    end of 122–3
    energy and light 8–9
    formation 3, 6–8
    photosphere 18
    rotation rate 35
    temperature 103
supercritical liquids 73
Super-Earths 85
superionic ice 72
supernovae 3, 51, 52–3
surface of planets 71–5
Swedenborg, Emmanuel 31–2

T

telescopes 24
    ALMA 26–30
    use of the transit method 80
temperature 6, 96–106
    categorization of planets 83
    critical 73
    identification on exoplanets 82
    and wavelength 18–19
terminator 98–100
TESS space telescope 80
thermodynamics 20
thorium 116, 124
tide-locked planets 90–1
Titan, clouds 106, 107
T_{rad} (radiating temperature) 102–3
transit depth spectroscopy 82
transit method 80–1
Trapezium cluster 10–11
Trappist-1 63, 77, 89
    habitability of planets
        112–13, 120
    orbital frequencies 92–4
tsunamis 37
T-Tauri stars 3–4, 7
    luminosity 56
    observations 25
turbulence 36
TW Hydrae 27

U

ultraviolet 15, 21
    and habitability 109
Universe
    age 50
    birth (Big Bang) 1
    knowledge of 128–9
Upper Scorpius star-forming
    region 29
uranium 116, 124
Uranus
    clouds 106
    formation 5–6

Uranus (*cont.*)
  gas accretion 46
  ice 64

V

Vega 11
Venus
  atmosphere 72, 118
  formation 9–10
  greenhouse effect 105
  habitability 108, 112
visible light astronomy 14–15
volatility 58–62
volcanos 70–1, 117, 119
Vonnegut, Kurt 72

W

WASP-12 77

water (H_2O) 72–3
  in CI chondrites 67–8
  formation 4
  and habitability 109, 112
wavelength 16–17, 21
waves, transporting angular
      momentum 36–7
white dwarf stars 50, 123
Wien–Boltzmann Displacement
      Law 18

X

X-rays 15

Y

young stellar objects
      22–6

SOCIAL MEDIA
Very Short Introduction

Join our community

www.oup.com/vsi

- Join us online at the official Very Short Introductions **Facebook** page.
- Access the thoughts and musings of our authors with our online **blog**.
- Sign up for our monthly **e-newsletter** to receive information on all new titles publishing that month.
- Browse the full range of Very Short Introductions online.
- Read **extracts** from the Introductions for free.
- Visit our library of **Reading Guides**. These guides, written by our expert authors will help you to question again, why you think what you think.
- If you are a teacher or lecturer you can order inspection copies quickly and simply via our website.

ONLINE CATALOGUE
A Very Short Introduction

Our online catalogue is designed to make it easy to find your ideal Very Short Introduction. View the entire collection by subject area, watch author videos, read sample chapters, and download reading guides.

http://fds.oup.com/www.oup.co.uk/general/vsi/index.html